OPTIMAL DESIGN OF PROCESS EQUIPMENT

ELLIS HORWOOD SERIES IN ANALYTICAL CHEMISTRY

Series Editors: Dr. R. A. CHALMERS and Dr. Mary MASSON
University of Aberdeen

HANDBOOK OF PRACTICAL ORGANIC MICROANALYSIS
S. BANCE
ORGANIC TRACE ANALYSIS
K. BEYERMANN
FOUNDATIONS OF CHEMICAL ANALYSIS
O. BUDEVSKY
INORGANIC REACTION CHEMISTRY Volume 1: Systematic Chemical Separation
D. T. BURNS, A. TOWNSEND and A. G. CATCHPOLE
INORGANIC REACTION CHEMISTRY Reactions of the Elements and their Compounds, Volume 2, Part A: Alkali Metals to Nitrogen, Volume 2, Part B: Osmium to Zirconium
D. T. BURNS, A. TOWNSHEND and A. H. CARTER
IMPROVED HOLLOW CATHODS LAMPS FOR ATOMIC SPECTROSCOPY
S. CAROLI
LABORATORY HANDBOOK OF PAPER AND THIN LAYER CHROMATOGRAPHY
J. GASPARIC and J. CHURACEK
ORGANIC ANALYSIS USING ATOMIC ABSORPTION SPECTROMETRY
SAAD S. M. HASSAN
HANDBOOK OF ORGANIC REAGENTS IN INORGANIC CHEMISTRY
Z. HOLZBECHER, L. DIVIS, M. KRAL, L. SUCHA and F. VLACIL
GENERAL HANDBOOK OF ON-LINE PROCESS ANALYSERS
DAVID HUSKINS
QUALITY MEASURING INSTRUMENTS IN ON-LINE PROCESS ANALYSIS
DAVID HUSKINS
ANALYTICAL APPLICATION OF COMPLEX EQUILIBRIA
J. INCZEDY
PARTICLE SIZE ANALYSIS
Z. K. JELINEK
OPERATIONAL AMPLIFIERS IN CHEMICAL INSTRUMENTATION
R. KALVODA
METHODS OF PROTEIN ANALYSIS
Edited by I. KERESE
HANDBOOK OF CHEMICAL EQUILIBRIA IN ANALYTICAL CHEMISTRY
S. KORTLY and L. SUCHA
ATLAS OF METAL-LIGANG EQUILIBRIA IN AQUEOUS SOLUTION
J. KRAGTEN
STATISTICAL THEORY AND METHODOLOGY OF TRACE ANALYSIS
C. LITEANU and I. RICA
SPECTROPHOTOMETRIC DETERMINATION OF ELEMENTS
Z. MARCZENKO
LABORATORY HANDBOOK OF CHROMATOGRAPHIC AND ALLIED METHODS
O. MIKES
STATISTICS FOR ANALYTICAL CHEMISTRY
J. N. MILLER and J. C. MILLER
SEPARATION AND PRECONCENTRATION METHODS IN INORGANIC TRACE ANALYSIS
J. MINCZEWSKI, J. CHWASTOWSKA and R. DYBCZYNSKI
HANDBOOK OF ANALYSIS OF ORGANIC SOLVENTS
V. SEDIVEC and J. FLEK
ANALYSIS WITH ION-SELECTIVE ELECTRODES
J. VESELY, D. WEISS and K. STULIK
MUTA GENICITY TESTING IN ENVIRONMENTAL POLLUTION CONTROL
F. K. ZIMMERMAN

John Elvey

OPTIMAL DESIGN OF PROCESS EQUIPMENT

K. URBANIEC
Associate Professor
Research Institute for Organization of Machine Industry
Warsaw, Poland

Translation Editor:
Dr. C. McDERMOTT
Department of Chemical Engineering, University of Birmingham

ELLIS HORWOOD LIMITED
Publishers · Chichester

Halsted Press: a division of
JOHN WILEY & SONS
New York · Chichester · Brisbane · Toronto

WYDAWNICTWA NAUKOWO-TECHNICZNE
Warsaw · Poland

English Edition first published in 1986 in coedition between
ELLIS HORWOOD LIMITED
Market Cross House, Cooper Street, Chichester, West Sussex, PO19 1EB, England
and
WYDAWNICTWA NAUKOWO-TECHNICZNE
Warsaw, Poland

Distributors:
Australia, New Zealand
Jacaranda-Wiley Ltd., Jacaranda Press,
John WILEY & SONS INC.,
G.P.O. Box 859, Brisbane, Queensland 4001, Australia
Canada:
JOHN WILEY & SONS CANADA LIMITED
22 Worcester Road, Rexdale, Ontario, Canada.
Europe, Africa:
JOHN WILEY & SONS LIMITED
Baffins Lane, Chichester, West Sussex, England.
North and South America and the rest of the world:
Halsted Press: a division of
JOHN WILEY & SONS
605 Third Avenue, New York, N. Y. 10158 U.S.A.

British Library Cataloguing in Publication Data
Urbaniec, K.
Optimal design of process equipment.
1. Manufacturing processes 3. Industrial equipment — Design and construction
I. Title II. Optymalizacja w projektowaniu aparatury procesowej, *English*
670.42 TS183

ISBN 0-85312-509-0 (Ellis Horwood Limited)
ISBN 0-470-20248-3 (Halsted Press)

Printed in Poland

Table of contents

Preface

Processes used in the chemical, food or pharmaceutical industries comprise a finite number of unit operations. These operations such as mixing, heating, filtration, evaporation and many others are performed in specialized equipment units. The design and construction of each unit is determined by the requirements of the specific unit operation rather than by the type of process employed. Consequently, the traditional notions of chemical equipment, refrigeration equipment, etc., are replaced in this book by the more general notion of process equipment.

The book consists of three parts. Chapters 1 and 2 are devoted to a review of relevant technological problems and to methodological foundations of equipment optimization. Chapters 3 through 6 are an introduction to design optimization and a review of optimum finding and other theoretical methods. In Chapters 7 through 11, characteristic optimization problems of process equipment are discussed and illustrated by numerous examples of design optimization for the equipment used in the chemical, sugar and power industries.

The design of a process plant should be based on the choice of:

– the best process,
– the most suitable plant structure,
– the optimal parameters of the process,
– the best designs of the plant equipment.

The three problems at the top of the above list are well known in process engineering and there is an extensive literature covering these and related topics, e.g. [3, 10, 28, 32, 55, 68]. However, there is still much to do in this area and only a small fraction of the accumulated theory has actually been implemented but, over the years, some systematic progress is evident. On the other hand, the last problem, which can be placed somewhere between process engineering and machine designing, has received little attention.

The subject of this book is closely related to problems of equipment design. In general engineering design, optimization methods are widely accepted especially in finding designs to which can be attributed least energy consumption, smallest weight, lowest cost, etc. It is precisely this type of application of optimization methods that has drawn the attention of this author. Thus, neither the problems of selection of equipment structure nor the questions of optimal control are considered here.

Optimization methods cannot easily be applied in practical design, as both familiarity with the technological subject and skill in using a variety of mathematical tools are necessary. It is not of much help if a mathematician takes part in problem formulation because, as a rule, the verification of the problem statement and of the usefulness of the optimization results are closely related to the engineer's responsibility for the final design.

The aim of this book is to help the reader in preparing to use optimization methods in the design of process equipment. The chapters devoted to the theoretical fundamentals review the most important topics required for design optimization and demonstrate their interrelation when applied to process equipment.

The theoretical treatment is supplemented by illustrative examples related to engineering practice. The final chapters of the book provide a review of optimization problems associated with selected groups of commonly used equipment. A modest volume like this, however, may only provide some simple, brief technical examples. More complex optimization problems can be found in the pertinent literature [8, 10, 14, 21, 22, 25, 28, 29, 52, 58, 73]. The appendices given at the end of the book review some major topics from appropriate branches of mathematics.

Such an approach is aimed at providing the reader with a convenient starting point for private study of the subject by assembling relevant information from a variety of sources. At the same time, duplication is avoided of the extensive literature on optimization theory [5, 6, 15, 16, 19, 26, 48, 60, 61, 90, 91] and its applications, available elsewhere [22, 28, 52, 62, 73].

The choice of contents and the proportioning of individual chapters is based on the author's experience gained when lecturing students of Warsaw Technical University and also engineers associated with the Polish machine building industry. As regards the examples, some of them are taken from real research or design problems solved by the author for the chemical and sugar equipment building industry. Other examples are taken from various publications.

Some introductory remarks should also be made concerning the connection between the subject of this book and computerized design calculations. The use of optimization methods does not necessarily require the computer however, certain computations may be so laborious that to automate these becomes almost imperative.

The reader of this book need not be skilled in computer programming and is free to study that subject independently if necessary.

The book is intended for readers who are interested in optimization while dealing, in their work or studies, with the lay-out or with the design of process equipment. The book can thus be useful to designers employed in engineering offices or R & D organizations in the process equipment industry. It should also prove useful for university courses in process and equipment engineering.

Major symbols

The list given below provides major symbols appearing in Chapters 2–7. Consistency in the symbols throughout the remaining Chapters cannot be observed because the problems are highly diversified; the symbols used there are explained as they appear.

X – decision space,
$\boldsymbol{x}$ – vector of decision variables,
$\boldsymbol{x}_j$ – jth vector of decision variables,
$\boldsymbol{x}^{(i)}$ – vector found in ith iteration,
x_i – ith decision variable,
$\boldsymbol{x}^{\mathrm{opt}}$ – optimal vector of decision variables,
$\boldsymbol{p}$ – vector of parameters,
V – set of feasible solutions,
f – objective function,
∇f – gradient of the objective function,
g_j – function appearing in jth equality constraint $g_j(\boldsymbol{x}) = 0$,
h_k – function appearing in kth inequality constraint $h_k(\boldsymbol{x}) \leqslant 0$,
f_a – augmented objective function (in the Lagrange multiplier method),
λ_j – Lagrange multiplier corresponding to jth equality constraint.

CHAPTER 1

Design problems related to process equipment

1.1. REVIEW OF TECHNOLOGICAL AND ECONOMIC PROBLEMS

The unit operations performed in individual parts of a process plant may be divided into the following groups:

– mechanical operations, e.g. classification, feeding and mixing of solid particles, grinding;

– hydromechanical operations, e.g. fluid flow, mixing of liquids, gases and suspensions, separation of mixtures (sedimentation, filtration, centrifuging), purification of gases from particulates;

– thermal processes, e.g. heating, cooling, evaporation, condensation;

– mass transfer processes, e.g. solubilization of solids, crystallization, extraction, absorption, desorption, drying;

– chemical reactions.

Most operations may be either batch or continuous. A continuous operation requires that a steady flow of materials is maintained through the process: the substrates must be fed continuously and the products continuously removed. A batch operation is initiated after the substrates have been supplied. When the operation is finished, the products are removed and the process may be repeated.

The present-day trends of industrial development, towards increased plant capacity and process intensification, create a number of technological problems related to unit operations, such as:

– equipment units larger in size;

– higher values of the parameters (e.g. pressure, temperature) responsible for the degree of difficulty of operations;

– need for automatic control;

– requirements of improved quality of the product, such as uniform feeding rate of solids, very low final moisture content of dried materials, etc.

These problems determine the paths of development of the equipment design, namely:

– increased dimensions;

– larger wall thickness, use of materials with improved strength properties and higher heat and corrosion resistances;

– tendency to give preference to continuous rather than batch processes;

– provision in the design for some desirable control properties of equipment units;

– when relying on well known designs, improving their quality with respect to process capacity and possibly at the minimum associated cost.

Design work is aimed, among other things, at determining the dimensions required for correct and safe operation together with the resulting consumption of energy and materials. Initially, only the most important (usually input and output) process parameters are known. It may thus become necessary to determine other parameters which should also be accounted for in the course of design, e.g. rated pressure, cooling water flow rate, etc. This can be done by using the mathematical description of the process. The mathematical description usually reflects such fundamental physical relationships as mass balance, energy balance, equilibrium conditions and kinetic equations. One can also use the description of phenomena occurring in the structure of the apparatus, e.g. stress distribution or wear at the exposed surfaces. In addition, it may become necessary to account for certain constraints resulting from the rules of engineering practice or mandatory codes; the constraints may be related to safety, ergonomics, attractive appearance as well as to the rules of economic calculations, the principles of cost analysis, the requirements imposed on cost or some other economic indices.

The majority of design problems have more than one and occasionally a number of feasible solutions. A unique design is then determined by a decision aimed at choosing just one of many feasible variants. In practice, decisions of this kind are usually taken on the basis of engineering experience or intuition, but it may also be possible to rely on some scientific basis. Optimization methods assist with the latter approach and make it possible for the decisions to be not only technologically feasible, but also optimal with respect to a certain criterion.

In design problems, process equipment may be evaluated using, for example, certain indices related to control properties, mass and cost of the equipment, or the indices expressing process capacity. The application of optimization methods thus enables one to design the equipment in conformity with the main trends of technological development, with the quality factor taken into account.

The possibilities of equipment design optimization should be considered bearing in mind that the design work is usually performed in several consecutive steps, namely design concept, initial design and final (detailed) design. The greatest potential to influence design quality is contained within the first two steps, which include, among other decisions, the determination of the main dimensions of the apparatus and of the main process parameters. It is in the initial steps of the design work that optimization methods are most rewarding.

1.2 PRACTICAL MEANING OF EQUIPMENT OPTIMIZATION

The suitability of optimization methods as design tools is based on their potential to solve problems that arise from interrelated technological and economic requirements. The fundamental economic requirement usually imposed upon process equipment design is that unit operations should be performed at a low total cost (i.e. combined capital and operating cost). This requirement may in some cases be expressed in terms of effectiveness measured by some cost-related index. If the cost or the effectiveness index can be exactly expressed in a mathematical form, then it is possible to use optimization methods and to determine the solution which can be attributed the lowest cost or the most advantageous effectiveness index.

In some cases, the method of cost calculation for a given equipment unit may be difficult to establish, because of the necessity to account for the role of the unit as a component in a processing plant. Although the plant is usually also evaluated in terms of cost or effectiveness, such an evaluation may be associated with the necessity to account for various factors which are independent of individual equipment units. Examples of such factors are: raw material utilization or energy consumption in the process. Eventually, economic calculations cannot be regarded as universally conclusive evaluation measures if for no other reason than because it is frequently necessary to evaluate processing plants from the point of view of their influence on the environment, control characteristics or other economically indefinable features.

If an evaluation measure is unambiguously defined for a processing plant, then it is possible to exactly determine the corresponding measure, using decomposition, for the equipment units in the plant. Decomposition is nowadays a specialized field with its own theory and computation techniques described elsewhere [27, 57]. Technically, decomposition can be applied only after the mathematical model of the entire plant has been developed, which makes its use difficult. Actually, incomplete data, management problems and other practical reasons often make it impossible to formulate the mathematical model of the plant and even to establish the relations between design objectives at different levels (i.e. plant – equipment units – elements of the equipment). This is characteristic of modern design methods which involve the use of catalogues which describe pre-designed equipment units and their elements.

The case of interconnected design problems, where the evaluation criteria are established separately and the problems are solved independently, are also to be expected.

In process equipment optimization the following criteria of equal rank are in use:

– structural,
– process-dependent,
– process/economic.

The most frequently used structural criterion is the mass or the volume of the material needed to construct the equipment unit or its part. Evidently, the evaluation based on mass or volume is incomplete as it ignores such substantial utilization factors as annual operating time, energy consumption etc. On the other hand, this criterion is certainly not as primitive as it may look at the first glance. Its importance results from the fact that in large-scale processing plants, the dimensions of equipment units are large and the units are bulky. If the mass of material in large units is reduced, then other economies follow, as the manufacturing, transportation and assembly costs are reduced. Sometimes, it becomes also possible to reduce the costs of supporting structures and buildings.

Other structural indices can be formulated depending on the specific use of the equipment.

The process-dependent criteria are much more diversified as they depend on the kind of unit operation performed. Among the most universal indices the consumption of power needed to perform the operation occupies a prominent place. As in the case of mass or volume, the importance of power consumption increases in large scale processes, because power becomes a substantial cost item there.

Among the process-dependent criteria, appear also those indices related to the automatic control of equipment operation. Criteria of this kind usually express the smoothness of operation, being defined on the basis of concepts of mathematical statistics (like variance or mean square deviation) attributed to important process parameters. Such criteria are widely used in optimal control problems and may be helpful in design applications, if it is known that the equipment design may affect the smoothness of equipment operation. Using optimization, one is able to arrive at a design which secures the best smoothness (best in the sense of the criterion used). For example, one may seek such a design of a mixer of loose materials as to secure minimal fluctuations of the concentration of components of the final mixture (see Chapter 3, Example 3.4). Such problems usually require the mechanisms of transient processes to be accounted for, which necessitates the application of control theory. For this reason, problems of this kind are not discussed in this book.

The combined process/economic criteria and the sense of optimality they imply have already been mentioned at the beginning of this Chapter. It should be mentioned that there are many reasons for controversy related to their use. First of all, economic calculation methods may be not generally applicable. The prices, and even the rules of cost calculation may change during the planned period of operation; this may be extremely difficult to anticipate while designing specific equipment. Although it does not exclude the possibility of optimization, it brings about a modification of the sense of optimality (see Chapter 3 for discussion).

Finally, it frequently happens that the application of a certain process/economic criterion creates the danger of a conflict with the economic evaluation methods, at least with some types of plant in which the equipment is used. Particular care is thus recommended when formulating the criteria of equipment evaluation – especially when it is impossible to analyse plant decomposition. Eventually, the process/economic criteria should be regarded as an evaluation tool, the application of which is desirable but not always practicable.

The evaluation methods used in equipment design reviewed above assume only one criterion. This is the simplest and the most frequently used evaluation technique. However, it may happen that design evaluation refers to several criteria simultaneously. The resulting problem becomes one of multicriterion optimization. When relying on the existing theoretical foundations, unique solutions to such problems can be obtained in special cases only. As practical experience from applications of multicriterion optimization to process equipment design is still limited, this special area is not discussed here.

1.3 EXAMPLE OF OPTIMAL DESIGN

Let us consider now a design concept of a pipeline for hydraulic transport of crushed solid material of known physical properties. At an initial design stage, the wear problems and the capital cost of the pipeline can be deliberately ignored; it can also be assumed that there are no limitations on the amount of water used in the plant. The required mass flow of solid material is G_s and the transportation distance is l. For the sake of pump life, the concentration of solid particles in water suspension must not exceed the limit value of ξ_l.

The pipeline may be defined by

– pipe diameter d,

– suspension flow velocity c,

– cost of the power consumed by the pump.

The mass balance of solid material can be expressed by the equation

$$V_s = \frac{G_s}{\rho_s} \tag{1.1}$$

where: V_s – volume flow of the material and ρ_s – material density.

Suspension flow velocity together with pipe diameter determine volume flow rate of the suspension

$$V = \frac{\pi d^2}{4} c \tag{1.2}$$

Thus the concentration of solid material in the suspension is

$$\xi = \frac{V_s}{V} \tag{1.3}$$

It follows from the problem formulation that the concentration is limited according to inequality

$$\xi \leqslant \xi_l \tag{1.4}$$

An energy balance on the material transported leads to the equation defining the pumping power N required to overcome the pressure drop Δp accompanying suspension flow

$$N = V\Delta p \tag{1.5}$$

while pressure drop is determined by an empirical relationship [35]

$$\Delta p = \frac{\rho c^2}{2} \frac{l}{d} \left[\lambda + 0.282\xi \left(\frac{\rho_s}{\rho} - 1 \right) B^{1/3} \left(\frac{gd}{c^2} \right)^{4/3} (1 + 43\xi^4) \right] \tag{1.6}$$

where: λ – friction coefficient of the pipe wall, B – a constant characterizing the material transported, g – acceleration due to gravity, ρ – suspension density.

The energy cost can be expressed by the unit cost index (cost per unit length of the pipeline and unit mass of material transported)

$$C_e = \frac{N\varkappa_e}{lG_s\eta} \tag{1.7}$$

where: $\varkappa_e$ – cost per unit of power, η – efficiency of the pumping set (pump and motor).

If the values of limit concentration, material constants, unit energy cost and pumping efficiency are known, then it is possible for any decision defined by a pair of values d and c to make the necessary substitutions into equations (1.1)–(1.7). However, there are many feasible decisions which satisfy inequality (1.4). For a certain set of values of these coefficients, several feasible decisions are given in Table 1.1. Possible design concepts differ in power costs, because volume flow V and pressure drop Δp in formula (1.5) vary depending on pipe diameter d and flow velocity c. Thus, varying power consumption makes the cost according to formula (1.7) vary. At certain specific d and c values, though, there is a minimum of energy consumption resulting from a compromise between the role of water as an agent decreasing the pressure drop and its role as a ballast increasing energy consumption in the transportation process. Recognizing energy cost as the essential evaluation criterion, one may accept as optimal the figure given at the bottom of the last column of Table 1.1. In addition, Fig. 1.1 shows a number of possible solutions and the importance of constraint (1.4).

Table 1.1

Unit cost index of energy consumed in hydraulic transport of a solid material with the density $\rho_s = 3300$ kg/m^3 and characteristic constant $B = 603$, at mass flow $G_s = 19.3$ kg/s; $\varkappa_e = 0.002$ £/kWh, $\eta = 0.7$; $\lambda = 0.016$. Limit value of the concentration of solid material in the suspension is $\xi_l = 0.4$

Internal pipe diameter d, mm	Flow velocity c, m/s	Unit cost index C_e, 10^{-2} £/t km
85.3	3.0	1.45
85.3	5.0	2.10
104.0	2.0	1.69
129.0	4.0	2.17
159.5	1.5	1.83
159.5	4.5	2.48
104.0	2.8	1.36

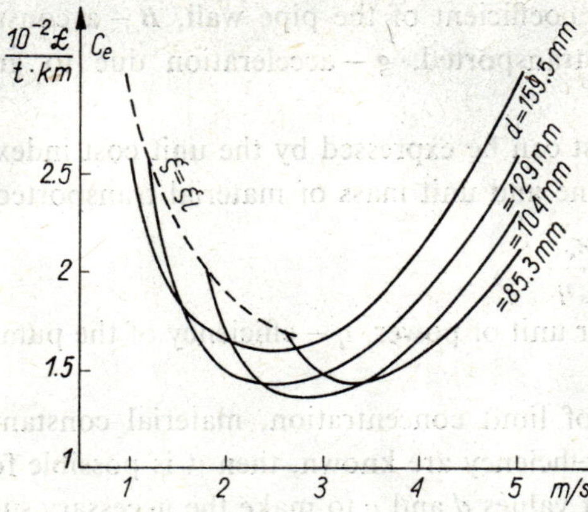

Fig. 1.1 Unit cost index of energy consumed in hydraulic transport as a function of pipe diameter and suspension flow velocity. The constraint is violated in the region above the dashed curve.

CHAPTER 2

Philosophy of optimal design

2.1 MORPHOLOGY OF DESIGN

As with other components of engineering knowledge, optimization methods can be used more or less skilfully, thus giving more or less satisfactory results. The need to apply them or their efficiency is not universal, if for no other reason than because their use may require an increased workload and prolonged duration of the design process.

The chances of obtaining benefit depend on a skilful introduction of optimization methods into the decision-making process aimed at the determination of designs [2, 34, 52]. Generally, the design process should be aimed at determining designs which satisfy all the necessary technological, economic and other conditions. These conditions result from existing necessities, although they may not be explicitly given. Hence it appears that the initial step should consist in problem identification which involves, among other things, defining the technological and economic conditions imposed upon the design. At the same time, the design principle should be chosen.

The designing of process equipment usually follows the scheme shown in Fig. 2.1. Its characteristic part is the initial design stage at which the main features of the object under consideration are determined. Other data needed to manufacture the object are determined at the detailed design stage. Next, building and testing of the prototype makes it possible to check the design correctness; if the result is unsatisfactory, it may be necessary to repeat at least some of the procedures involved in initial and detailed design. Once a desirable result has been obtained, the documentation concluding the design process can be provided.

Each of the stages shown in the scheme may have its own internal structure.

This is exemplified by the left-hand part of Fig. 2.1, where a frequently used, traditional scheme of initial design is visualised. Its characteristic feature is that the proposed solutions are arbitrarily formulated on the basis of engineering experience, intuition or model solutions. Next, engineering calculations are performed to check whether or not the solution advanced satisfies the necessary conditions and limitations. Examples of such calculations are comparison of stress value under extreme operating

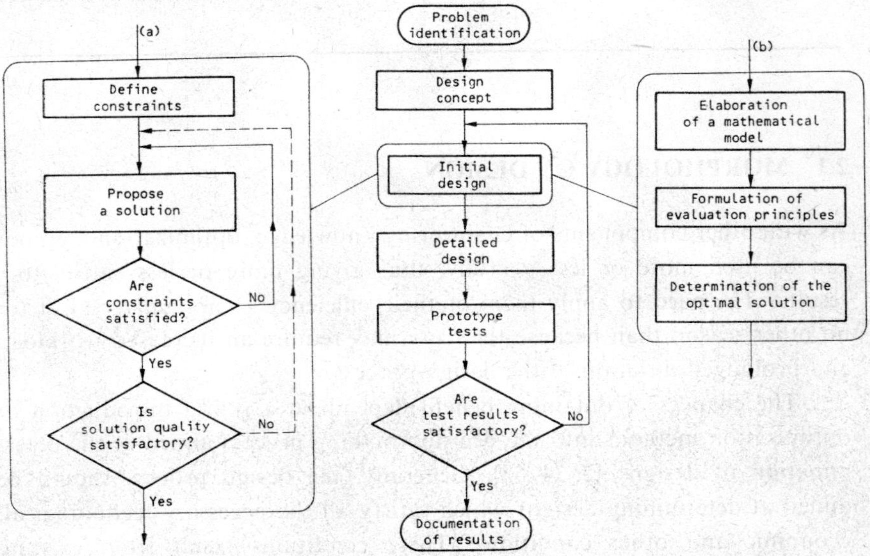

Fig. 2.1 Main stages of the design process (center) and the initial design procedure based on arbitrary decisions (a) or on optimal decisions (b).

conditions with the maximum allowable stress, the operating temperature with the highest allowable temperature, or the cost with its limiting value, etc. If the check results are unsatisfactory, then an alternative solution should be proposed; acceptable results, on the other hand, are an indication that the objectives of the initial design have been attained. In some cases, this scheme additionally includes the examination of the quality of solutions by comparison with the existing models; if the result is unsatisfactory, another solution is to be sought.

The above scheme is thus based on iterative examinations of proposed solutions and may be described as a *trial and error design procedure*, or *design analysis*. An alternative procedure, shown in the right-hand part of Fig. 2.1, is called a *synthetic design procedure*, or *design synthesis*. It is based on a properly devised mathematical model which may cover conditions and limitations as well as the examination of design quality. The most important feature of the design synthesis is that the solution can exactly be determined as a result of operations performed on the mathematical model. The quality factor can be reflected by the fact that in those operations, optimization methods are applied for determination of a solution which can be attributed the highest quality. This improves the chance for the final design to be the best of all available choices. It is the basic property of design synthesis which in that respect differs from the trial and error method which gives no assurance that the best result will be found. Even if the thus determined initial design includes a quality evaluation shown in Fig. 2.1 by a dotted line and a satisfactory solution has been found, it is generally not known whether or not better solutions do exist. The doubts cannot be eliminated, even if several satisfactory results are compared and the best one is chosen (except for the trivial case when the number of available choices is finite and all of them can be evaluated and compared).

It should be pointed out that the initial design procedure has been considered here only as an example of the application of optimization methods in design. These methods can be applied to various problems of detailed design or even to prototype testing. Examples of such problems are: determination of the most suitable set of tolerances in the parts of a mechanism, determination of the cheapest program of prototype tests, etc.

2.2 MATHEMATICAL MODELLING

The modelling of a certain object consists in creating its analogue – a *model* – the properties of which are related to the properties of the original. It is commonly believed that the connection between the properties is based on the fact that both the model and the original are physical objects and that their properties can be measured (e.g. dimension, stress or temperature). Actually, the connection between the model and the original may be purely conceptual, as the suitability of the model depends only on exact knowledge of the connection. Thus, if only certain definite rules are observed, an economic system can be modelled by a hydraulic device, or a mechanical system – by an electrical analogue.

A mathematical model is created as an abstract analogue to an object under consideration. This is a mathematical structure which may include equations, inequalities, functions, etc. While in physical modelling the necessary data about the original are gathered by model experiments and subsequent transformation of their results, mathematical modelling relies on calculations, also called numerical experiments.

Engineers have always used the essential principle of mathematical modelling in engineering calculations. However, 'true' modelling will usually require computer-aided calculations in which detailed mathematical descriptions of the problem can be used instead of traditional approximative formulae. Another attribute of modelling is the application of proper calculation methods leading to high accuracy of numerical results. Also, the use of ample empirical data in conjunction with other theoretical relationships may prove very helpful.

It should be emphasized that mathematical modelling may be used in solving various technological problems. As a result, a variety of meanings may be attached to that concept. Also, the methods used to arrive at a model and the mathematical apparatus employed in modelling may vary with the case at hand. For example, in control problems the model should reflect the relationships between certain quantities describing the condition of the object, defined as 'output variables' – and other quantities, also describing the condition of the object, defined as 'input variables'. Account should also be taken of the influence of two other groups of quantities known as 'control variables' and 'disturbances'. Such problems may require modelling of steady states (static models) or transient states (dynamic models) of the object. Characteristically, the numerical experiments are aimed at determining the course of phenomena associated with the object.

The idea of model investigation is thus to imitate the experiments with the real object at which measuring equipment has been installed with the aim to observe input, output and control variables as well as the disturbances. The advantage of the mathematical model lies in the fact that it can be easily applied making it also possible to anticipate the course of events, while real experiments only enable one to record these. This is an invaluable feature in the investigations of breakdowns or emergency conditions in the object considered. However, it should be remembered that the current state of knowledge may sometimes prove insufficient to describe the object (e.g. on the basis of the relationships given in the literature). In that case, the formulation of a mathematical model may require carrying out of simple experiments aimed at the verification and accuracy of the model or even at the determination of coefficients in theoretical relationships (so-called model identification).

Mathematical models used in design also have their own requirements which form the subject of Chapter 3. One particular feature consists in the fact that numerical experiments reflect the consequences of design decisions. Such numerical experiments can only be replaced by prototype tests. The main advantage of a mathematical model is the possibility of anticipating facts which otherwise can be investigated only by costly and time-consuming physical experiments. Again, a word of caution is in order in that the formulation of a model may require certain experiments aimed at the verification of theoretical relationships.

2.3 OPTIMIZATION

The term 'optimization' is commonly understood as the determination of the best – perhaps ideal – solution of a problem. In the rigorous meaning of the word, however, a condition is implied that the required solution is the best one under specific circumstances for the assumed evaluation criterion. In the field of equipment design, optimization should be understood as the determination of that design, among all those which satisfy the conditions of technological feasibility, which renders a maximum or minimum of some evaluation criterion or objective function.

The above statement should be treated as a brief characterization of the problem, rather than a definition. Let us compare it to a somewhat more comprehensive description given in [8]: "Optimization is a branch of science devoted to the methods of choosing optimal solutions to any actions, e.g. technological or economic ones. In particular, optimization of structural design includes the problems of choosing shape parameters and physical properties of structures. The choice consists in determining a solution which is optimal in the sense of the criterion formulated above.

By the shape of the structure we shall understand not only its dimensions but also other properties like strength or strain properties of the material. Accordingly, instead of 'optimization' the term 'proportioning' can also be used".

It should be remembered that all formulations based on abstract concepts are ambiguous unless definitions of those concepts are introduced. Consequently, the present author has deliberately made no attempt to define such concepts as 'solution', 'technological feasibility', etc. to circumvent the terminological and methodological controversies still appearing in the literature. It may therefore be useful to say that the range of optimization problems considered in this book does not include the problem of choosing the optimal working principle of the equipment, although this problem may arise in the initial stages of designing. Usually, problem identification (see Section 2.1) results in determining one or several comparable working principles and the mathematical optimization tools presented in the following chapters are suited to precisely such a case.

In order to gain better understanding of the above statement, let us consider two examples of design cases involving the choice of the working principle of the equipment.

In the first case, a solution is sought for the problem of transferring to the environment a certain amount of waste heat contained in a process liquid. For this purpose the equipment types shown in Fig. 2.2 can be used. These are: heat exchanger cooled with river water, air cooler and wet

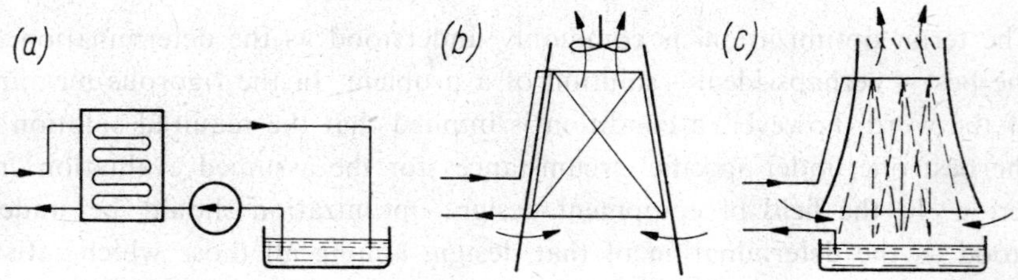

Fig. 2.2 Choosing a design concept of cooling equipment: (a) water cooler, (b) air cooler, (c) wet cooling tower.

cooling tower. These three types are not entirely comparable as their performances and shortcomings are unequally distributed among the fields of economics, environment protection, ease of operation, etc. The choice of a working principle to enable us to solve the problem in an optimal way is thus beyond the scope of this book. Well within its scope is, however, the problem of optimal design of any equipment type, e.g. with respect to the mass of the material used or to the over-all cost including capital and running expenditures. The same applies to the concomitant optimization of equipment design and process parameters related to the coolant.

In the second case, a design is required of a shelf-type condenser to be operated at definite process parameter levels. Three condenser variants are feasible which differ by shelf geometry (Fig. 2.3). As the manufacturing and operating problems are nearly the same here – the variants are fully comparable and choosing one of them falls within the scope of the problems under consideration.

In practice, there is a great deal of misunderstanding about the criteria used in design optimization. This seems to result from the fact that optimality, as it is commonly and imprecisely understood, and its mathematical meaning are frequently confused. The usefulness of optimization methods, however, is by no means limited to the situations when an ideal optimization criterion can be formulated. Their basic meaning as an engineering tool is to enable one to determine the relations between unequivocally stated design objectives and the decisions needed to attain them. The doubts about an evaluation criterion can usually be resolved by determining

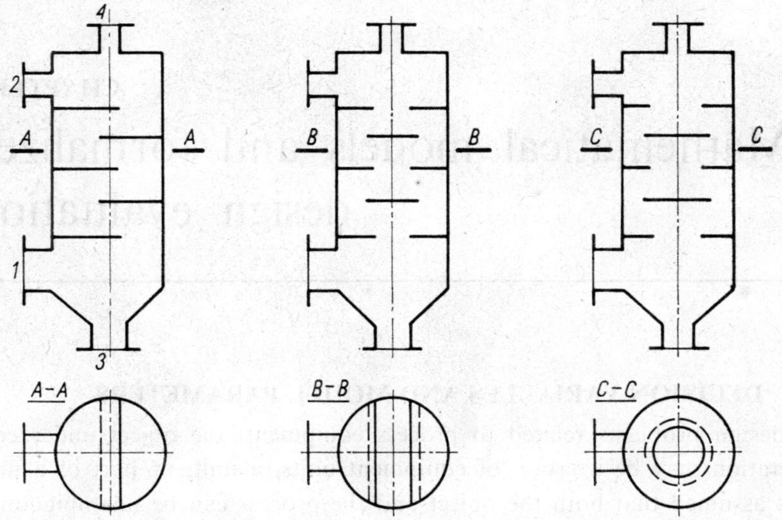

Fig. 2.3 Choosing shelf design in a condenser. 1 – vapour inlet, 2 – cooling water inlet, 3 – water outlet, 4 – incondensible gases outlet.

a set of decisions close to the optimal one (with respect to that criterion), or by finding several decisions, each being optimal with respect to a differently formulated criterion. Accordingly, optimization always allows one to prepare the final decision better than intuitive design methods which are nearly always subject to inconsistencies inherent in the reasoning.

CHAPTER 3

Mathematical models and formalized design evaluation

3.1 DECISION VARIABLES AND MODEL PARAMETERS

In design problems related to process equipment, the object under consideration may be a group of equipment units, a unit, or part of a unit. It is assumed that both the object and the process can be unambiguously described with the aid of certain quantities of a technological or economic nature. The quantities to be determined in the course of solving a design problem are formally treated as unknown variables; they are called *decision variables*. The remaining quantities which are not subject to the designer's choice are treated as constants; they are called *parameters*.

Development of a mathematical model of the design problem consists in identifying the decision variables and parameters and in formulating all constraints and other important interrelations. The main difficulty in model development is a necessity to satisfy two requirements:

– correct description of the technological aspects of the problem in as great a detail as the technological needs imply,

– mathematical validity of the model in a form suited to practical possibilities of solving the problem by analytical or numerical methods.

3.2 MATHEMATICAL MODEL OF THE DESIGN OBJECT

The mathematical modelling of the design object may be difficult particularly if no similar cases are available in the literature. A proper model is usually obtained by an iterative trial-and-error modelling process based on technological expertise which enables one to investigate the meaning and the conditions of the design problem. Nevertheless, it may be useful to remember a simple methodological rule that mathematical modelling always consists in transforming specific technological facts into abstract formulae. This transformation can be made easier, if the design object is analysed in technological terms to which definite mathematical analogues correspond.

As stated above, there is a correspondence between the quantities present in the description of a design problem, and decision variables and parameters appearing in the mathematical model. As concerns their practical meaning, these quantities can be classed into two groups: structural quantities and process quantities. In equipment design problems, the decision variables usually, but not always, turn out to be structural variables, while the parameters usually have the meaning of process quantities. The opposite is the case, however, for process engineering problems.

Structural quantities characterize the structure of the object, e.g. its shape and the way its parts fit together, together with the physical properties of the object, like dimensions and cross section areas and some object-related effects e.g. stress build-up, wear, etc.

The process quantities, on the other hand, characterize the condition and physicochemical properties of the substances participating in the process; also, some effects of which the process consists, like heat and mass exchanges, chemical reactions, flows, etc.

The above division is purely illustrative and not entirely exact, as the quantities important to the design of the specific item may also be vital to the process.

The interrelations between structural and process quantities are reflected by relationships between decision variables and model parameters. Essentially, they arise from:

– dependences between object properties, e.g. dimensional constraints, relations between mass and dimensions;

– phenomena occurring in the object structure, e.g. stress, wear of parts;

– relations between process quantities, e.g. thermodynamic equations, mass and energy balances;

– basic phenomena occurring in the process, e.g. mass and heat transfer, chemical reactions;

– technological conditions and economic and other requirements, e.g. safety rules, cost calculation principles, etc.

Technological correctness and proper consideration of detail do not have direct counterparts in the mathematical model. Both factors, however, affect the formal properties of the model. The amount of detail, for example, contributes to the number of quantities under consideration; its increase makes the elaboration and utilization of the model more difficult. Thus, taking into consideration variables which are unnecessary from the point of view of model validity is clearly a mistake. It should also be borne in mind that a simple problem description allows for the checking of the modelling results by intuitive interpretation.

Example 3.1

A mathematical model is to be developed for a finned plate shown in Fig. 3.1 which transfers heat to a boiling liquid of known properties. The model should make it possible to determine optimal fin dimensions for heat transfer effectiveness.

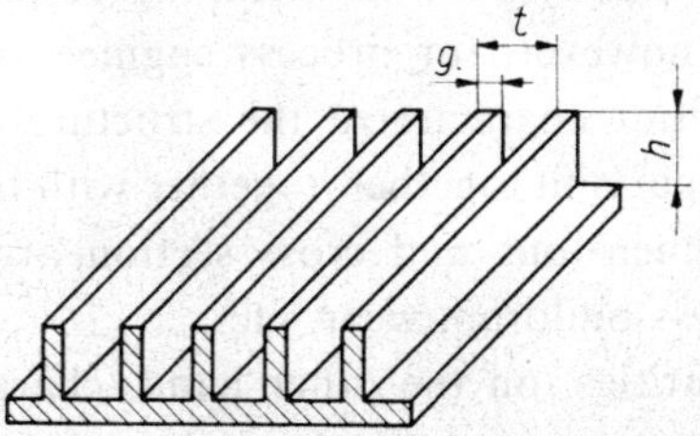

Fig. 3.1 A plate with finned surface.

Structural quantities characterizing the object under consideration are: fin height h, thickness g and pitch t; also, thermal conductivity of fin material λ. The heat transfer process may be characterized by the following quantities: amount of heat transferred in unit time from unit area of plate surface Q, temperature difference between plate and liquid ΔT, heat transfer coefficient α.

The amount of the heat transferred results from heat transfer from fins (Q_1) and from the remaining plate surface (Q_2)

$$Q = Q_1 + Q_2 \tag{3.1}$$

Heat transferred from fins

$$Q_1 = \alpha \Delta T \eta F_1 \tag{3.2}$$

where: η – fin efficiency, F_1 – fin surface area. If the distances between fins are large enough, then the coefficient α is independent of fin geometry and only depends on temperature difference:

$$\alpha = A \Delta T^{2.33} \tag{3.3}$$

where A – empirical constant depending only on liquid properties. Thus α need not be treated as an independent process quantity.

Fin efficiency depends on fin geometry, thermal conductivity of the material and some process quantities. Introducing the notation

$$m = \sqrt{\frac{2\alpha}{\lambda g}} \tag{3.4}$$

$$x = mh \tag{3.5}$$

$$y = 1.75 - 0.078x \tag{3.6}$$

fin efficiency in a boiling liquid can be expressed as

$$\eta = \frac{\tanh(xy)}{xy} \tag{3.7}$$

It follows from the geometrical relation for the finned surface that on a unit surface of the plate

$$F_1 = 2\frac{h}{t} \tag{3.8}$$

By inserting (3.3)–(3.8) to (3.2), an equation is obtained which links Q_1 with all the structural and process quantities.

Similarly, the amount of heat transferred from the remaining plate surface can be expressed as

$$Q_2 = \alpha \Delta T F_2 \tag{3.9}$$

where

$$F_2 = 1 - \frac{g}{t} \tag{3.10}$$

The above formulae enable one to express total amount of heat Q as a function of other structural and process quantities. Thus, Q need not be treated as an independent process variable.

For manufacturing reasons, additional conditions concerning fin height, thickness and pitch must be taken into account:

$$g \geqslant g_{\min} \tag{3.11}$$

$$t \geqslant t_{\min} \tag{3.12}$$

$$h_{\max} \geqslant h \geqslant h_{\min} \tag{3.13}$$

where: $g_{\min}$, $t_{\min}$, $h_{\max}$, $h_{\min}$ are limit dimensions.

Another condition of a clearly geometrical meaning must always be satisfied, viz.

$$t > g \tag{3.14}$$

The relations given above create the required mathematical model. Decision variables are h, g, t and parameters ΔT, A, λ. Equations (3.1) to (3.10), expressing the influence of the decision variables and parameters upon process quantity Q, enable one to quantitatively investigate heat transfer effects. Decision variables are bound by inequality constraints (3.11)–(3.14).

3.3 DECOMPOSITION OF HIGHER-LEVEL MODELS

As already mentioned in Section 1.2, the way in which the equipment is modelled and evaluated should be compatible with evaluation principles adopted for higher-level systems. In particular this applies to economic feasibility evaluations. Hence, it may be necessary to consider the problem of decomposition of a system which includes the equipment to be designed.

A rigorous decomposition of the mathematical model of a high-level system is hardly ever possible but is not always necessary. There are certain ways of checking what effects related to the design object should be accounted for in the evaluation. The reasoning is based on the fact that there are mass and energy flows between the high-level installation and the design object. These flows can be described using n process quantities at the object inlet and outlet. As these are bounded by m equations in the mathematical model, only $n-m$ of them may vary independently. Among the design data, s process quantities may be given. Two basically different cases can then be identified:

(1) $s = n-m$

This is the case of a total 'separation' of the object from the high-level system. The designer is free to choose structural quantities and only those process quantities which reflect the processes inside the object. The other case is when decision evaluation applies to effects related to the design object only, viz.

(2) $s < n-m$

Here, a certain freedom of decision exists concerning the connections between the object and the system. The design decision affects not only the design object, but also the associated plant and these effects should be accounted for in the evaluation.

Example 3.2

The problem is to determine design evaluation principles for a fired heater for an oil-processing plant (Fig. 3.2). The plant design is evaluated with respect to the total annual cost.

The heater heats up the oil at a mass flow G from temperature t_1 to temperature t_2. Inlet pressure is p_1. The oil should be prevented from boiling in the heater, which implies that the outlet pressure must not fall below a certain value $p_{2\,\min}$ which is much lower than p_1.

The relations between the heater and the plant are characterized by five independent process quantities G, t_1, t_2, p_1, p_2. Four of them have been unambiguously defined in the design problem statement and an

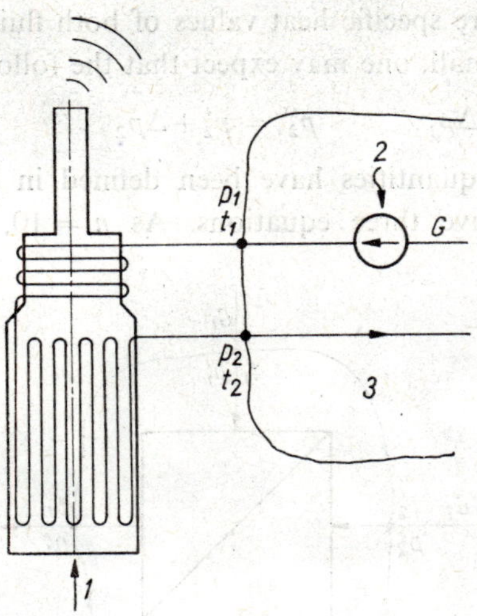

Fig. 3.2 A fired heater connected to a plant. 1 – fuel, 2 – energy, 3 – plant.

inequality constraint including p_2 has been given. Thus, $n = 5$, $m = 0$, $s = 4$, which implies that $s < n-m$.

According to case (2), the heater design evaluation should include the effect on the plant of design value of p_2. A proper effect measure is the annual cost C_e of energy needed for pumping the oil through the heater. As the total annual cost of the heater is approximatively equal to the sum of fixed cost C_f and fuel cost C_l, the heater evaluation should be based on the combined annual cost

$$C = C_e + C_f + C_l$$

Example 3.3

The problem is to determine evaluation principles for the design of a heat exchanger operated in a process plant, Fig. 3.3. No phase change occurs in any of the liquids flowing through the exchanger. The plant design is evaluated with regard to the sum of fixed cost and operating cost.

The heat exchanger is required to cool a liquid at mass flow G_1 from temperature t_1' to temperature t_1''. The mass flow of the cooling agent is G_2 and its initial temperature is t_2'. Inlet pressure values in both fluids are p_1' and p_2', while pressure drops cannot exceed relatively small values of Δp_1 and Δp_2.

The relations between the exchanger and the plant are described by ten process quantities G_1, t_1', t_1'', p_1', p_1'', G_2, t_2', t_2'', p_2', p_2''. The heat balance is reflected by equation

$$G_1 C_1 (t_1' - t_1'') = G_2 C_2 (t_2' - t_2'')$$

where: C_1 and C_2 are specific heat values of both fluids. As the allowable pressure drops are small, one may expect that the following relations hold:

$$p_1'' = p_1' + \Delta p_1, \qquad p_2'' = p_2' + \Delta p_2$$

Accordingly, seven quantities have been defined in the design problem formulation, to derive three equations. As $n = 10$, $m = 3$ and $s = 7$,

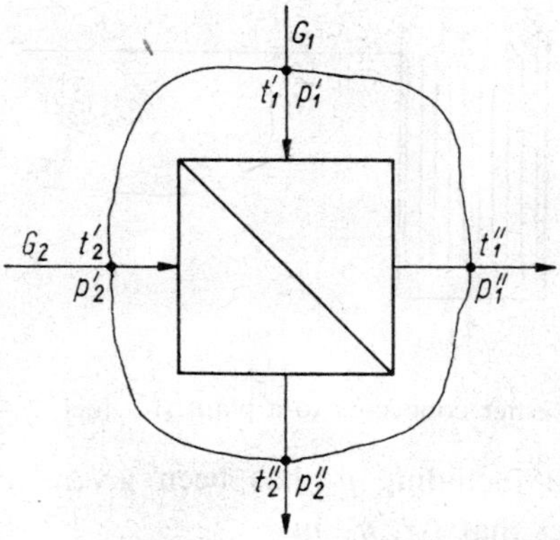

Fig. 3.3 A heat exchanger in a process plant.

the relation $s = n - m$ is valid. Hence, the value of heat exchanger cost may serve as a satisfactory basis for the exchanger design evaluation. As this case at hand is in agreement with case (1), it should be sufficient to consider the exchanger mass alone.

3.4 FORMAL PROPERTIES OF MATHEMATICAL MODEL

Before going on to the further sections of this Chapter the reader is advised to study Appendices 1 and 2. This should facilitate the use of terminology and notation characteristic of set theory and vector calculus which are applied in our treatment.

The mathematical correctness of the model, as well as its suitability to effectively determine the solutions to design problems, both depend in the first place on the properties of decision variables, parameters and the constraint system.

Different variants of the problem solution can be considered as elements of a certain mathematical space called the *decision space*. Then, the determination of values of all decision variables is interpreted as the determination of an element in that space.

To define the decision space mathematically, it is necessary to investigate the properties of all decision variables. According to the practical meaning of each variable, it can be qualified as a real variable, random variable, real-valued function etc. Almost always, the decision space has the properties of a linear topological space. Most often, if the decision variables are real ones, it is simply a multidimensional Euclidean space.

A handy convention is that an element (or a point) in a space may be defined by specifying real numbers which are its coordinates. These numbers are sometimes treated as the components of a vector with its origin at the centre of the coordinate system. It should, however, be remembered that this would be a special case of the decision space, because decision variables do not always have to be real ones, they could be functions or logical variables as well. Then, decision space would be non-euclidean and it might prove difficult to find a convenient mathematical definition of that space. This may happen, for example, when the design problem involves searching for the structure of the object under consideration. It may sometimes turn out to be completely impossible to formalize problems of that type. (If the number of feasible structures is small, however, they can be treated as variants requiring individual consideration).

A concept analogous to the decision space is the *parameter space*. Defining it and investigating its structure is not always necessary or useful, because in many problems the parameters are simply known numbers. Then, the designer has to deal with a simple point in the parameter space and the suitability of this concept is limited. It may happen, however, that in the first attempt to identify the problem before the design work starts, lack of essential data is discovered. The definition of the parameter space may then prove to be a major part of the initial analysis leading to the completion of basic information. As examples of design cases in which the concept of parameter space may be of significance, mention may be made of the design of an equipment unit for variable operating conditions or the designing of a series of units.

A *system of constraints* in a mathematical model usually comprises inequalities and equations. The inequalities generally express the requirements imposed upon decision variables or their functions.

A variety of equation types may occur in the models, for example:

– algebraic and transcendental,
– ordinary differential,
– partial differential,
– integral and differential-integral.

Algebraic and transcendental equations are most commonly used in mathematical descriptions of design objects. This type of equations is applied to express such relations as those between object properties or between process quantities, as well as relations characterizing steady state phenomena in lumped-parameter systems or design rules of an empirical nature. Ordinary differential equations are generally useful in describing transient states in lumped-parameter systems or steady states in distributed-parameter systems. Characteristically, the equations must be completed by initial or boundary conditions.

Partial differential equations are applied to describe transient states of distributed-parameter systems or steady states of multidimensional distributed parameter systems. Complete specification involves, of course, initial conditions in the first case and boundary conditions in the second.

As for integral and differential-integral equations, these are rather seldom encountered in design problems.

Some types of constraint systems are particularly awkward because of lack of universally applicable and effective methods (analytical or numerical) for their solution. This applies in the first place to:

- systems of nonlinear equations in many variables,
- boundary problems for the systems of ordinary differential equations,
- partial differential equations.

As an important complement to the mathematical model, one should take account of the physical meaning and qualitative aspects of the relationships and phenomena described. This can make it easier to devise algorithms for their solution, or enable one to find good starting points for iterative calculations.

3.5 CLASSIFICATION OF MODELS

Certain properties of mathematical models affect the principles of evaluation of design decisions and also the choice of methods of finding optimal solutions. Hence, it is of some practical importance to recognize those properties while formulating the model.

Mathematical models used in design can be divided into three groups which differ with respect to the manner in which input information is defined:

- models with complete information (deterministic),
- models with random-type information (probabilistic),
- models with incomplete information.

In deterministic models, the values of all decision variables and parameters can be unambiguously determined. This ensures a complete description of the design object and an unambiguous formulation of all constraints. The formulation of such a description is practicable only for the case where the operating conditions are precisely known a priori and the implementation of design (manufacturing and putting into operation) will proceed in a strictly controlled manner.

Probabilistic models include at least one decision variable or parameter with the properties of a random variable. This is the description applicable to design problems in which uncertainty is allowed with respect to some process or structural quantities or with respect to the manufacturing of the object. Probabilistic models require that statistical data or other information are available for defining the distributions of all random variables. As an example of a random variable, one can name a dimension to be realized within a definite tolerance interval (gaussian distribution of the variable can be assumed in that case).

Models with incomplete information include decision variables and parameters exhibiting the properties of random variables (or stochastic processes) of unknown distributions. Alternatively, decision variables or parameters may be known only to belong to a given set or random variables or to assume real values from a given set. This can happen in design problems which are formulated on the basis of prognostic or estimated data. As an example of a random variable of unknown distribution, mention may be made of the estimated price of a raw material some years hence.

Deterministic models are applied in design much more frequently than are probabilistic models or models with incomplete information. This results from the fact that in the majority of commonly used or standard engineering calculation methods, a deterministic problem description is assumed. It should be mentioned that fully proven theoretical rules of modelling and optimization generally exist for deterministic problems only. Therefore, in order to avoid the difficulties in solution, problems of an essentially probabilistic nature are often modelled according to the deterministic scheme. Noteworthy, however, is the fact that models with incomplete information make it possible to analyze and to solve certain problems which cannot be properly formulated in deterministic terms.

Example 3.4

A mathematical model is to be developed for a continuous mixer of loose materials which makes a mixture of two materials (1 and 2) delivered by two feeders of known characteristics. The model should make it possible to determine the most advantageous dynamic properties of the mixer.

A characteristic feature of the modern design of feeders of loose materials is that allowance must be made for flow fluctuations which are rather substantial. Instantaneous mass flow of materials can be treated as random processes $m_1(t)$ and $m_2(t)$. Concentration of material 1 in the mixture at the mixer outlet is then

$$c_1(t) = \frac{m_1(t)}{m_1(t)+m_2(t)} \tag{3.15}$$

Concentration fluctuations are compensated for in the mixer because material particles remain there for a longer time. The mixing process can be regarded as random with respect to the particle holdup time, u, which can only be described by the probability density function $h(u)$ satisfying condition

$$\int_0^{\infty} h(u)\,du = 1 \tag{3.16}$$

This function is determined by mixer design. Usually, no analytical form of $h(u)$ in structural variables can be given, although a mean value or variance of the retention time can be estimated for different designs.

The concentration of material 1 in the mixture at the mixer outlet can be expressed by equation

$$c_0(t) = \int_0^{t} h(u)\,c_1(t-u)\,du \tag{3.17}$$

The dynamic properties of the mixer can be regarded as advantageous, if the variance of c_0 (variance is the measure of fluctuations) is small. Taking into account known characteristics of feeders and the relations (3.15) and (3.17), one can draw conclusions about the most desirable properties of the function $h(u)$ and, subsequently, about the most advantageous mixer design. Obviously, before any calculations are performed, the mathematical description of process quantities m_1 and m_2 should be determined.

3.6 DESIGN EVALUATION IN DETERMINISTIC PROBLEMS

If the design object is described by a deterministic mathematical model, the evaluation criterion can be usually formulated as a figure of merit; mathematically, this is a function defined in decision space. Most often, if the decision space is of finite dimension then the figure of merit is a real-valued function of the decision variables; it is known as objective function, criterion function or return function. The figure of merit assigns a number to each element of the decision space which reflects the quality of the design represented by this element. For example, if the equipment design is evaluated with respect to its mass, then the figure of merit should be formulated as a relation between the mass and decision variables (dimensions, cross sections, material densities). Usually, a design evaluated with respect to its mass is regarded as better, if the value of the figure of merit is lower. Generally, it is also possible to adopt a principle that the higher the value of the figure of merit, the better is the design. This is the case of evaluation with respect to equipment capacity, capacity-to-cost ratio, etc.

Depending on the meaning of the criterion, the optimal solution is characterized by minimum or maximum value of the measure of merit. Mathematically, it is unnecessary to distinguish between minimum and maximum, it is merely the fact that an extremum is attained that matters. Every index of a specific technological meaning can be multiplied by -1 or subtracted from a large number, thus transforming the problem of searching for maximum into a minimum search, or conversely.

Example 3.5

An objective function to be used in the evaluation of a plate container of a given volume (Fig. 3.4) is to be defined.

The container is cylinder-shaped; it is manufactured by welding together the jacket and two circular heads. The designer has to choose diameter D and height H of the cylinder.

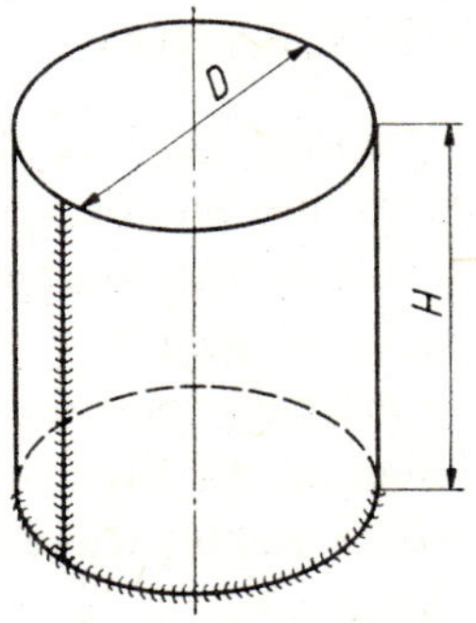

Fig. 3.4 A cylindrical plate container.

The container should be designed for the minimum quantity of plate consumed, which is equivalent to attaining minimum surface area (plate thickness is constant). The objective function whose minimum corresponds to the optimal solution can be written down as

$$f_1(D, H) = \pi DH + 2\frac{\pi D^2}{4}$$

As dividing f_1 by π does not change minimum location, a simpler function form can be written, viz.

$$f_2(D, H) = DH + \frac{D^2}{2}$$

Of course, finding a minimum of f_1 and f_2 is equivalent to finding maximum of function

$$f_3(D, H) = -\left(DH + \frac{D^2}{2}\right)$$

The example given above demonstrates how a simple formal scheme of decision evaluation can be introduced in deterministic cases. Denoting an arbitrary element of decision space by $\boldsymbol{x}$ and a given element of parameter space by $\boldsymbol{p}$, the evaluation can be based on function value $f(\boldsymbol{x}, \boldsymbol{p})$. The optimal solution corresponds to such an element $\boldsymbol{x}^{\text{opt}}$ of decision space that

$$f(\boldsymbol{x}^{\text{opt}}, \boldsymbol{p}) = \operatorname*{extr}_{\boldsymbol{x}} f(\boldsymbol{x}, \boldsymbol{p})$$

In the abbreviated form based on the fact that parameter values are known, the search for the optimal solution can also be symbolized as

$$f(\boldsymbol{x}) \to \operatorname*{extr}_{\boldsymbol{x}} f$$

3.7 DESIGN EVALUATION IN PROBLEMS WITH PROBABILISTIC OR INCOMPLETE INFORMATION

Probabilistic or incomplete information in a design problem description may introduce new components into the formalism of solution evaluation. It then becomes necessary to rely on concepts taken from probability theory and mathematical statistics, such as expected value, variance, etc. Even if it follows from the practical meaning of the problem that it is necessary to evaluate the design on the basis of a certain definite measure of merit depending on decision variables $\boldsymbol{x}$ and parameters $\boldsymbol{p}$, the meaning of optimality is by no means clear.

Even in the simplest problem of a probabilistic nature, with just one component of $\boldsymbol{x}$ or $\boldsymbol{p}$ being a random variable, it can be immediately seen that the formulation

$$f(\boldsymbol{x}, \boldsymbol{p}) \to \operatorname*{extr}_{\boldsymbol{x}}$$

cannot be attributed an unambiguous meaning, because an extremum point attained at a certain value of the random variable need not be extremum at another value of that variable.

Two different techniques can be used to eliminate the uncertainty introduced by random variables, namely:

(1) searching for extremum of the measure of merit with expected values inserted instead of the random variables themselves,

(2) searching for the extremum of the expected value of the measure of merit.

Finally, the basic idea of the optimality criterion becomes, in both cases, identical with that of the deterministic problem: the optimal solution corresponds to the extremum of a function defined on decision space.

To illustrate the above remarks, let us consider the formulation of a measure of merit in a problem featuring two-dimensional space X and one-dimensional space P, while p is a random variable which may take values between $+1$ and -1, and the probability of different values is characterized by the density function $\psi(p)$ such that $\int_{-1}^{+1} \psi(p)dp = 1$.

Let us assume that for a given parameter value p, the quality of design decisions can be expressed by the objective function

$$f(\boldsymbol{x}, p) = x_1 p^2 + x_2 p - x_1 x_2$$

Then, for $p = -1$

$$f(\boldsymbol{x}, -1) = x_1 - x_2 - x_1 x_2$$

while for $p = 0$

$$f(\boldsymbol{x}, 0) = -x_1 x_2$$

and for $p = +1$

$$f(\boldsymbol{x}, 1) = x_1 + x_2 - x_1 x_2$$

It can be easily seen that for each of the three parameter values, entirely different functions in variables x_1 and x_2 are obtained and thus the extremum points can be entirely different.

Taking into account the two possible optimality criteria formulated above, one can arrive in case (1) at

$$f(\boldsymbol{x}, \mathscr{E}(p)) \to \underset{x}{\text{extr}}$$

where $\mathscr{E}(p) = \int_{-1}^{+1} p\psi(p)\, dp$ is the expected value of p.

In case (2), the criterion becomes

$$\int_{-1}^{+1} f(\boldsymbol{x}, p)\psi(p)dp \to \underset{x}{\text{extr}}$$

and, after performing the integration with respect to p, an objective function is obtained which depends on x only.

In the second group of problems characterized by mathematical models including random variables of unknown distributions, it is impossible to directly apply the above criteria. Instead, one can proceed as follows [74]:

(1) after defining a set of deterministic problems which represent special (possibly extreme) cases of the general problem, the optimal solution in the deterministic sense is found for each problem;

(2) after evaluating all the solutions in the set defined in the previous step, optimal solution is found on the basis of game theory criteria (Fig. 3.5).

The principles of game theory can be applied to determine optimal course of action in a game, i.e. in a process which requires making sequen-

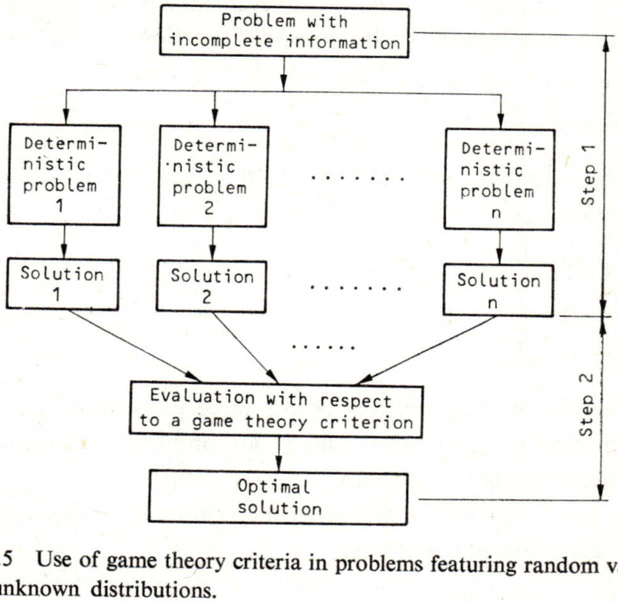

Fig. 3.5 Use of game theory criteria in problems featuring random variables with unknown distributions.

tial decisions under changing conditions (resulting from random factors or from the counteractions of an adversary). The decision can be evaluated according to such criteria as

– minimization of maximum losses (minimax criterion),

– minimization of mean losses.

The same criteria are applicable to situations which cannot be interpreted as 'game' but in which incomplete information should be accounted for in the decision making.

There exists an extensive literature devoted to game theory and the details can be found there [88]. Let us observe, however, that in the two-stage procedure mentioned above, the first step consists of solving a set of deterministic problems, in which the sense of optimality is realised by the attaining of the extremum of a function.

Example 3.6

Let us consider here the evaluation of the fired heater design under uncertainty conditions.

The heater discussed in Example 3.2 should be evaluated with respect to its annual cost C which is the sum of capital cost, fuel cost and cost of power required for oil pumping. In order to make it possible to evaluate the heater design, three economic parameters should be known: unit cost of heat-resistant steel c_s, unit cost of fuel c_f and unit cost of power c_e.

In the deterministic case, one should choose design $\boldsymbol{x}$ which corresponds to the minimum annual cost $C(\boldsymbol{x}, c_s, c_f, c_e)$. However, the expert's view is that during the heater life, fuel cost may vary between $c_f^{\min}$ and $c_f^{\max}$ and power cost between $c_e^{\min}$ and $c_e^{\max}$. Most probably, fuel cost dispersion will be considerable. On the other hand, steel cost can be strictly determined as it occurs only once at the time when the heater is manufactured.

Step (1) of the optimization procedure consists in finding quasi-optimal (i.e. minimizing function C) designs at six different combinations of fuel and power costs (Table 3.1). It is assumed that marginal costs are

Table 3.1

The principle of generating quasi-optimal designs (Example 3.6)

Case	Fuel cost	Energy cost	Optimal solution	Annual cost at optimum
1	$c_f^{\min}$	$c_e^{\min}$	$\boldsymbol{x}^{(1)}$	C_{11}
2	c_f^{m}	$c_e^{\min}$	$\boldsymbol{x}^{(2)}$	C_{22}
3	$c_f^{\max}$	$c_e^{\min}$	$\boldsymbol{x}^{(3)}$	C_{33}
4	$c_f^{\min}$	$c_e^{\max}$	$\boldsymbol{x}^{(4)}$	C_{44}
5	c_f^{m}	$c_e^{\max}$	$\boldsymbol{x}^{(5)}$	C_{55}
6	$c_f^{\max}$	$c_e^{\max}$	$\boldsymbol{x}^{(6)}$	C_{66}

investigated; additionally, mean fuel cost c_f^m is also considered as the dispersion of the expert's forecasts concerning fuel costs is particularly large.

At step (2), the annual cost is calculated for each quasi-optimal design at changed economical conditions (Table 3.2). If the minimax criterion is applied in choosing the optimal solution, then maximum annual cost C_i^{max} should be found in Table 3.2 for each economic situation investigated (i.e. for each line). The optimal design $\boldsymbol{x}^{(l)}$ corresponds to $C_l^{max} = \min_i C_i^{max}$.

Table 3.2

The principle of investigating quasi-optimal solutions and preparing the choice of the optimal solution with the aid of game theory criteria (Example 3.6)

Design	Case 1	2	3	4	5	6	Maximum cost	Mean cost
$\boldsymbol{x}^{(1)}$	C_{11}	C_{12}	C_{13}	C_{14}	C_{15}	C_{16}	C_1^{max}	C_1
$\boldsymbol{x}^{(2)}$	C_{21}	C_{22}	C_{23}	C_{24}	C_{25}	C_{26}	C_2^{max}	C_2
$\boldsymbol{x}^{(3)}$	C_{31}	C_{32}	C_{33}	C_{34}	C_{35}	C_{36}	C_3^{max}	C_3
$\boldsymbol{x}^{(4)}$	C_{41}	C_{42}	C_{43}	C_{44}	C_{45}	C_{46}	C_4^{max}	C_4
$\boldsymbol{x}^{(5)}$	C_{51}	C_{52}	C_{53}	C_{54}	C_{55}	C_{56}	C_5^{max}	C_5
$\boldsymbol{x}^{(6)}$	C_{61}	C_{62}	C_{63}	C_{64}	C_{65}	C_{66}	C_6^{max}	C_6

If a minimal mean loss criterion is applied, then one should calculate the mean arithmetic value of annual cost from all economic situations C_i, for each $\boldsymbol{x}^{(i)}$ in Table 3.2. The optimal design $\boldsymbol{x}^{(m)}$ corresponds to $C_m = \min_i C_i$.

CHAPTER 4

Mathematical properties of optimization problems

4.1 FORMAL APPROACH TO OPTIMIZATION PROBLEMS

The essential role of a mathematical model of a design object is to define decision space X and, within this space, to determine the set V of elements satisfying all constraints; V is the *set of feasible solutions.* A criterion for decision evaluation can be interpreted as function f defined in space X. A mathematical problem which repeatedly occurs in optimization using different evaluation approaches, is to search for the element $\boldsymbol{x}^{\text{opt}}$ in space X which belongs to set V and simultaneously minimizes function f

$$f(\boldsymbol{x}^{\text{opt}}) = \min_{\boldsymbol{x} \in V} f(\boldsymbol{x}) \tag{4.1}$$

Such problems can be divided into two groups depending on whether space X is finitely dimensional (e.g. Euclidean space, the elements of which can be treated as points in multidimensional space) or infinitely dimensional (function space whose elements are functions). In the first case problem (4.1) consists in searching for the extremum of a function of many variables; in the second case it is a variational problem. The reader should be warned, however, that with respect to variational problems, the notation used in (4.1) is rather vague and does not reflect the subtleties of a variety of problems belonging to variational calculus [10, 60].

Another distinctive feature of problem (4.1) is related to the properties of set V. If it is identical with space X or if it is an open set, which is equivalent to lack of constraints or to the presence of equality constraints only, then the problem is called classical. If V is not an open set, then the problem is called non-classical.

In the field of process equipment, optimization problems usually resolve themselves into searching for the extremum of a function. Below, a classification is given of problems of this kind, i.e.

$$f(\boldsymbol{x}^{\text{opt}}) = \min_{\boldsymbol{x} \in V} f(\boldsymbol{x}), \qquad V \subset X$$

(1) Unconstrained extremum

$$V = X$$

(2) Equality constrained extremum

$$V = \{x: g_i(x) = 0, \quad i = 1, 2, \ldots, n\}$$

(3) Mathematical programming problems

$$V = \{x: g_i(x) = 0, \quad i = 1, 2, \ldots, n; \quad h_j(x) \leqslant 0, \quad j = 1, 2, \ldots, m\}$$

where: X – Euclidean space; f, g_i, h_j – real-valued functions. Most practical problems are non-classical ones, which means they can be regarded as mathematical programming problems.

Henceforth, we will be dealing only with the problems of extremum of a function.

A common belief is that a good formulation of a problem is equivalent to half of its solution. When searching for the extremum of a function, this is certainly true as some properties of the detailed problem formulation may considerably affect the degree of difficulty of its solution; among others, the efficiency of certain computational methods may be impaired. Consequently, careful identification of the problem form should be treated as an important step. Quite often, such identification enables one to modify the mathematical formulae without changing the essence of the problem, thus making the problem easier to solve.

4.2 DECISION SPACE

If X is a Euclidean space, then all coordinates of point x representing an arbitrary decision may assume arbitrary values from certain continuous intervals. This is a *problem with continuous variables.*

If X is a set of discrete points in Euclidean space, then coordinates of x may assume values from certain discrete sets. This is a *problem with discrete variables.*

An intermediate case is also possible in which X has the property that some variables are continuous and others are discrete; this is a *mixed problem.*

It can generally be said that continuous problems are easier to solve, whereas the presence of discrete variables makes it more difficult. If possible, one should avoid discrete variables in the formulation of optimization problems; alternatively, an elimination technique may be tried. An example of the elimination of discrete variables by using the properties of the objective function is shown in Example 4.6. Another elimination technique is based on the principle of transforming the problem into a series of problems with continuous variables only, with the aid of a penalty function of a special type (*see* Section 5.3.3.2).

A most widely used method of solving a problem with discrete variables involves 'embedding' it into a problem with continuous variables. This is a two-step procedure, viz.

(1) the discrete nature of the variables is disregarded and an extremum of the objective function is found as if all variables were continuous (finding of apparent optimum);

(2) the neighbourhood of the apparent optimum is searched and true extremum is found. In the second step, the methods of sensitivity analysis can be used (*see* Section 6.2).

4.3 CONSTRAINT SYSTEM

The mathematical significance of the constraint system lies in the fact that it defines the set of feasible solutions V (set of all x's which satisfy the constraints) in decision space. The optimal solution must belong to that set. Before problem solving is attempted, the properties of set V should be carefully investigated for their possible consequences.

In the first place, one should be aware of the possibility that constraint properties may imply hedging the optimal solution with qualifications. For example, if random parameters appear in the constraints, then the boundary of the set of feasible solutions may be random. As a consequence, some points in the decision space may or may not satisfy the constraints, depending on parameter values. Should the optimal solution be chosen among such points, then it would become necessary to provide a practical interpretation of that fact, for example in the operation manual where appropriate qualifications would need to be included.

In the course of design calculations, one of the following techniques of transforming a random into a deterministic set has to be used:

(1) replace random parameters by their expected values,

(2) replace constraints relating to the functions of random variables by constraints related to the expected values of those functions,

(3) assume the acceptable (i.e. reasonably small) value of the probability of violation of constraints.

The most appropriate technique may be difficult to choose, particularly if some of the constraints reflect safety rules, wear conditions, etc. A more detailed discussion of these problems would require the use of probability theory [62]. It is assumed here that it is always possible to define a deterministic region of feasible solutions.

A region of feasible solutions may be simply connected or multiply connected. A simply connected region has such a property that two arbitrary points in it can always be connected by a continuous line which also belongs to this region. In a multiply connected region, this would be impossible. An example of such a region is the set of feasible solutions in a two-dimensional problem with two constraints:

$$-x_2+\frac{1}{6}x_1 \leqslant 0$$

$$-2\sin x_1+x_2 \leqslant 0$$

Graphical interpretation of this example is shown in Fig. 4.1.

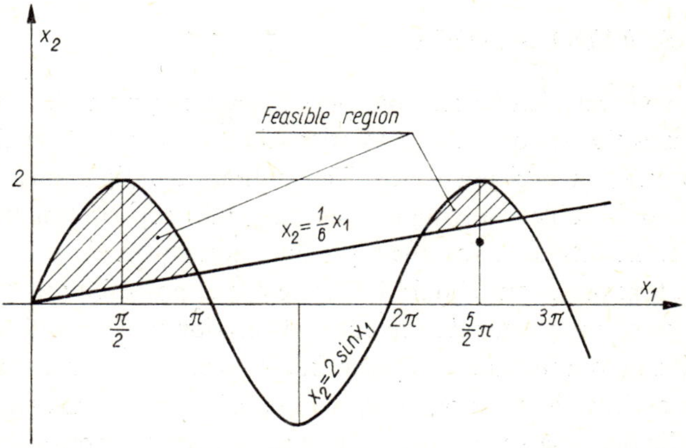

Fig. 4.1 Example of a two-dimensional, multiply connected region of feasible solutions.

It may happen in multidimensional problems that it would be difficult to establish whether or not the region of feasible solutions is simply connected. This implies the necessity to make allowance for a multiply connected region, which makes solution more difficult.

Example 4.1 (from [53])

The minimum of function

$$f(x_1, x_2) = (x_1-5)^2+(x_2+10)^2$$

is to be found in the region defined by inequalities

$$-2x_1+\frac{1}{30}x_1^2+x_2 \leqslant 0$$

$$-x_2 \leqslant 0$$

$$x_1(x_1-1)(x_1-15)(x_1-30) \leqslant 0$$

Suppose that the search after solution is performed by random trials (this type of methods will be discussed in Chapter 5). It is very likely that after a few trials some point will be found in the neighbourhood of (15, 0) with the best function value about 200. Minimal function value at point (1, 0), however, is 116. This point is located in that part of the feasible region which is very small in comparison to the other part, Fig. 4.2.

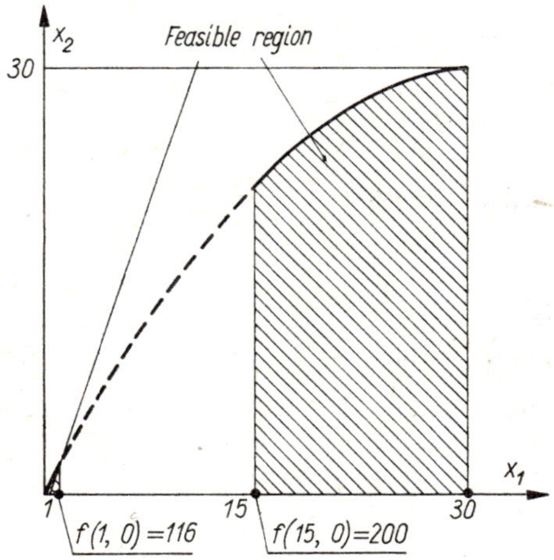

Fig. 4.2 Geometrical representation of the solution of Example 4.1.

Another important property of the region of feasible solutions is the convexity. A set is convex if two arbitrary points inside it can be connected by a segment of a straight line, while all points of that segment are also located inside the set. For example, if all constraints are linear, then the region of feasible solutions is convex. Most methods of finding the extremum of a function ensure finding the solution in convex regions only.

The simpler the constraint form and the lower the number of constraints, the easier is the recognition of important properties of the set of feasible solutions. It is always recommended to check whether the constraint system can be simplified by making use of its particular mathematical features. For example, an equality constraint of the form $g(\boldsymbol{x}) = 0$ containing variable x_k enables one to substitute $x_k = G(x_1, \ldots, x_{k-1}, \ldots \ldots, x_{k+1}, \ldots, x_n)$ and then to eliminate this variable from the problem completely. It should, however, be considered whether removing one variable and reducing the dimensionality of the problem is really advantageous, as the substitution may make the analytical form of other constraints and of the objective function more complicated.

Certain forms of inequality constraints enable one to transform the variables in such a way that the region of values of a new variable is infinite, which facilitates problem solving.

For a constraint of the type

$$a_i \leqslant x_i \leqslant b_i \tag{4.2}$$

where: a_i and b_i are constants, such an effect can be obtained by inserting one of the expressions

$$x_i = a_i + (b_i - a_i)\sin^2 y_i \tag{4.3}$$

$$x_i = a_i + (b_i - a_i)\frac{e^{y_i}}{e^{y_i} + e^{-y_i}} \tag{4.4}$$

Similarly, a constraint of the type

$$x_i \geqslant a_i$$

where a_i – constant,
can be eliminated by inserting

$$x_i = a_i + y_i^2 \tag{4.5}$$

or

$$x_i = a_i + e^{y_i} \tag{4.6}$$

All the transformations shown above ensure that x_i satisfies the original constraint for arbitrary y_i values. The number of variables in the transformed constraint remains unchanged.

The insertion technique can also be used to convert inequality constraints into equations; such a transformation is necessary if certain solution techniques are to be employed. This, however, causes the number of variables to increase by 1 at each insertion. For example, a constraint of the form $h(\boldsymbol{x}) \leqslant 0$ can be replaced by equation $h(\boldsymbol{x}) + z^2 = 0$, where z is a new variable.

It may happen that some inequality constraints in a mathematical model are inessential from the mathematical point of view. In other words, such constraints are redundant because other constraints are 'more active'. Such a situation is shown in Fig. 4.3. If redundant constraints are identified and eliminated before problem solving is initiated, one can avoid unnecessary calculations of the values of functions appearing in these constraints. Elimination of such constraints can be performed on the basis of analytical investigations or numerical experiments.

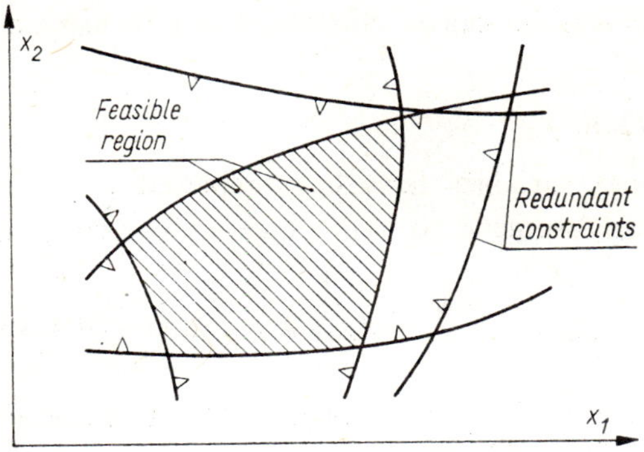

Fig. 4.3 Geometrical representation of a constraint system in two-dimensional space; two constraints are redundant. 'Edges' point at that part of the plane where the constraint is satisfied.

Example 4.2

Constraints are to be eliminated from the problem of minimization of function

$$f(x_1, x_2, x_3) = -x_1 x_2 x_3$$

subject to constraints

$$0 \leqslant x_1 \leqslant 42$$

$$0 \leqslant x_2 \leqslant 42$$

$$x_1 + x_2 - 88 \leqslant 0$$

$$0 \leqslant x_1 + 2x_1 + 2x_3 \leqslant 72$$

Firstly, it can be noted that the third constraint must be satisfied, if both preceding constraints are valid. The third constraint is thus redundant and can be disregarded in our subsequent reasoning.

Now if we perform a trigonometrical transformation

$$x_1 = 42\sin^2 y_1 \qquad x_2 = 42\sin^2 y_2$$

then the two initial inequalities can be eliminated. An analysis of the form of the last constraint results in the conclusion that another insertion would be appropriate:

$$x_1 + 2x_2 + 2x_3 = z$$

which is equivalent to

$$x_3 = \frac{1}{2}(z - x_1 - 2x_2)$$

Now, the last constraint can be eliminated too, by using a trigonometrical transformation

$$z = 72\sin^2 y_3$$

Once the original variables have been replaced by y_1, y_2 and y_3, the objective function assumes a form in which all four constraints are satisfied

$$f(y_1, y_2, y_3) = -\frac{42^2}{2}\sin^2 y_1 \sin^2 y_2 (72\sin^2 y_3 - 42\sin^2 y_1 - 84\sin^2 y_2)$$

Now, the unconstrained minimum of the above function can be found to yield optimal values $y_1^{opt}, y_2^{opt}, y_3^{opt}$. The corresponding $x_1^{opt}, x_2^{opt}, x_3^{opt}$ are obtained by simple substitution.

4.4 OBJECTIVE FUNCTION

Most optimization methods rely on a series of mathematical operations performed on the objective function. The suitability of a given method depends on whether or not the necessary operations are feasible. Particularly important can be such function properties as: continuity, existence of first derivatives and their continuity, existence of second derivatives, etc.

An important property of the objective function is whether the dependence on decision variables is linear or nonlinear. Linearity always makes the solution finding easier. In the case of nonlinear functions, other properties like convexity, separability and unimodality can also be of significance.

The convexity of a function means that function values determined by linear interpolation between two points are greater than true function values. Fig. 4.4 shows an example of a function in one variable; this function is either convex or concave depending on the interval considered. A useful mathematical interpretation of convexity is that the value determined by linear interpolation can always be regarded as upper estimate of the function value. In some numerical optimization methods, this property is used to set up a series of approximations to the solution.

The separability of a function means that the function can be represented as a sum of functions, each depending on one variable only:

$$f(\mathbf{x}) = \sum_{i=1}^{n} \phi_i(x_i) \tag{4.7}$$

The extremum of such a function can be found by successively finding the extremum of each of functions $\phi_1, \phi_2, \ldots, \phi_n$.

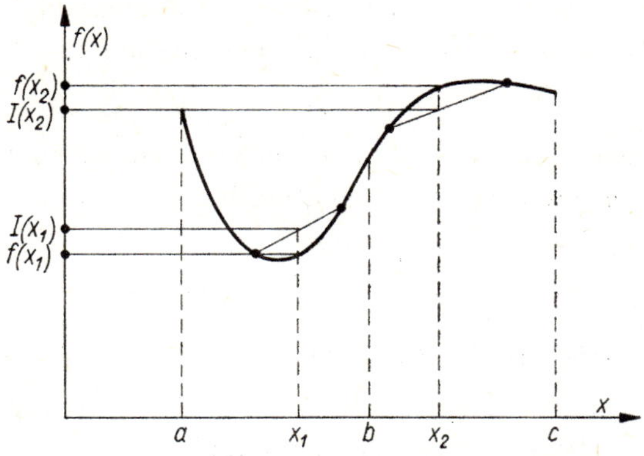

Fig. 4.4 Example of a function which is convex in interval $[a, b]$ and concave in $[b, c]$. $I(x)$ is function approximation obtained by linear interpolation.

Unimodality of a function with regard to a certain variable in a certain interval means that the interval under consideration can be split into two subintervals such that the function is strictly decreasing in one subinterval, and strictly increasing in the other, as shown in Fig. 4.5. A unimodal function has but one extremum in the interval considered; this makes it possible to use a wide class of efficient numerical methods of solution finding, other function properties being unimportant (the function can be discontinuous).

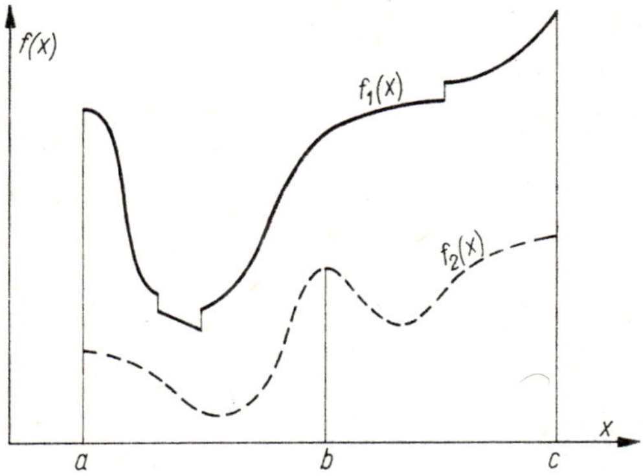

Fig. 4.5 Function $f_1(x)$ is unimodal in the entire interval $[a, c]$, while $f_2(x)$ is unimodal in each of the intervals $[a, b]$ and $[b, c]$.

Example 4.3

Function

$$f(x_1, x_2) = x_1^2 + 10x_1x_2 + 100x_2^2$$

is to be modified with the aim of obtaining a separable function. New variables can be introduced using following definitions:

$$z_1 = x_1 + ax_2 \qquad z_2 = x_2$$

where: a – constant (its value will be determined later). If the function is modified by introducing inverse relationships

$$x_2 = z_2 \qquad x_1 = z_1 - az_2$$

then we obtain

$$f(z_1, z_2) = z_1^2 + z_1 z_2(10-2a) + z_2^2(100-10a+a^2)$$

It can be easily seen that upon assuming $a = 5$, the term containing $z_1 z_2$ vanishes and the function becomes separable:

$$f(z_1, z_2) = z_1^2 + 75z_2^2$$

Example 4.4

The maximum of

$$f(x_1, x_2) = 3x_1^2 + 3x_2^4$$

is to be found subject to constraints

$$5x_1^2 + 2x_2^4 \leqslant 10$$

$$x_1^2 + 2x_2^4 \leqslant 4$$

Upon inserting $x_1^2 = y_1$ and $x_3^4 = y_2$, an equivalent problem is obtained

$$f(y_1, y_2) = 3y_1 + 3y_2 \to \max_{y_1, y_2}$$

$$5y_1 + 2y_2 \leqslant 10$$

$$y_1 + 2y_2 \leqslant 4$$

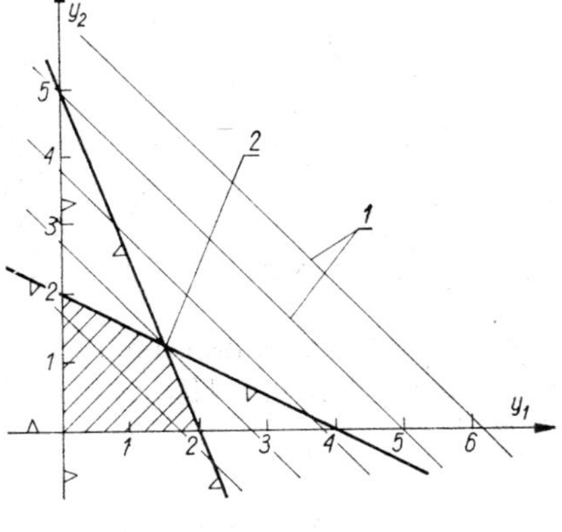

Fig. 4.6 Graphical representation of Example 4.4. 1 – contours of objective function, 2 – solution $y_1^{opt} = 3/2$, $y_2^{opt} = 5/4$.

$$y_1 \geqslant 0$$
$$y_2 \geqslant 0$$

This is a linear problem; its graphical interpretation and its solution are shown in Fig. 4.6. There are four solutions to the original problem, as

$$x_1^{\text{opt}} = \pm\sqrt{3/2} \qquad x_2^{\text{opt}} = \pm\sqrt[4]{5/4}$$

4.5 LOCAL AND GLOBAL EXTREMUM

A point at which a function attains the greatest or the least value from all function values in a set considered is called the *global* (absolute) *extremum point*. Optimization problems are usually focused on the global extremum.

Depending on the properties of the objective function or on the combination of function and constraint properties, at certain points in the feasible region the function value may be greater (or smaller) than in the neighbourhood of these points. Such points are called points of *local extremum* (Figs. 4.7 and 4.8).

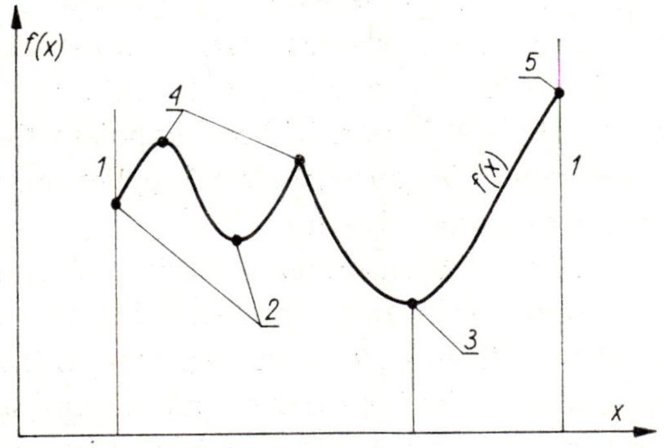

Fig. 4.7 Examples of extremum points of a function of one variable. 1 – constraint, 2 – local minimum, 3 – global minimum, 4 – local maximum, 5 – global maximum.

A popular technique used in extremum finding (in the absence of constraints) relies on investigations of points where first derivatives of the function vanish. This technique (discussed in more detail in Chapter 5) can only be used for continuously differentiable functions. In optimization problems, however, functions may appear that are not differentiable or are even discontinuous. Such functions may possess extremum points at discontinuities or where function derivatives are not defined or discontinuous. Such extremum points can be identified by investigating function values or values of the first derivatives at the neighbouring points. In Fig.

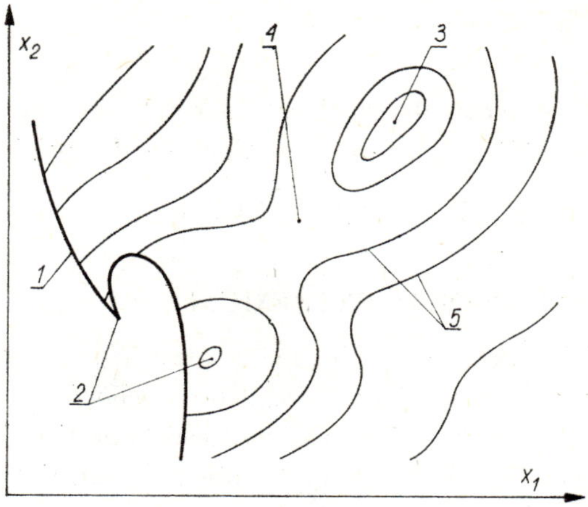

Fig. 4.8 Examples of extremum points of a function of two variables. 1 – constraint, 2 – local minimum, 3 – global minimum, 4 – saddle point, 5 – function contours.

4.9 the idea of such investigation is shown for a function of one variable. The test on function values is always concerned with three points: current point x_c and neighbouring points $x_c - \varepsilon$, $x_c + \varepsilon$. If function value at x_c is smaller (or greater) than at both neighbouring points, then x_c is a point of minimum (or maximum). Of course, practical application of this principle must be based on appropriately chosen ε value.

The other technique mentioned above consists of testing the signs of first derivatives at two points located at opposite sides of the current point. If the signs are different, there may be an extremum at the current point.

It should be added that both these extremum identification techniques are also good for constrained problems, providing that proper algorithms are constructed with the aim of selecting test points at the boundary of the

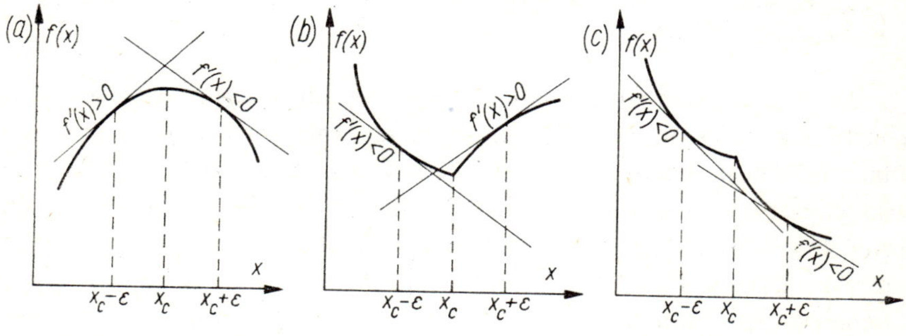

Fig. 4.9 Principle of extremum identification at points where first derivative vanishes (a) or is discontinuous (b, c).

feasible set. On the other hand, both techniques only find a point which may be an extremum point, without determining whether it is a local or global extremum. If several extremum points have been found, then it is by no means certain that no more such points exist. As a consequence, if these techniques are applied to extremum identification, then at least a rough initial diagnosis of the problem becomes necessary in order to gain some idea of the number and of the positions of local extremum points.

Actually, the above remarks are true for all numerical techniques of extremum identification. Various techniques are in use depending on the properties of the extremum finding algorithm.

4.6 PROBLEM DECOMPOSITION

In Chapter 2, model decomposition has been discussed as a tool for the correct formulation of optimization problems. As to the decomposition of mathematical optimization problems, this is an efficient way of simplifying the calculations necessary for problem solving. The basic idea consists of transforming a multidimensional problem into a series of problems in fewer dimensions which results in reduced computational effort. In the same way, variable elimination in discrete problems can be performed.

The decomposition technique is quite simple; it relies on drawing conclusions from examination of the properties of the objective function and constraints. This can be sometimes surprisingly efficient, as can be seen in the following examples.

Example 4.5 (from an unpublished report by J. Oderfeld)
Let us find the minimum of

$$f(x_1, x_2, x_3) = x_1 x_2 + \frac{x_1^2}{x_3}$$

where x_2 and x_3 are discrete variables:

$$x_2 \in \{2, 3, 4\} \qquad x_3 \in \{1, 2, 3\}$$

In this example, decomposition makes it possible to replace a three-dimensional problem by three one-dimensional problems. It can be easily seen that at constant values of x_1 and x_3, the function is linearly dependent on x_2, attaining its smallest value at $x_2^{\text{opt}} = 2$ for $x_1 > 0$ and $x_2^{\text{opt}} = 4$ for $x_1 < 0$. If the values of x_1 and x_2 are constant, then the dependence on x_3 can be easily established: the function attains its minimum at $x_3^{\text{opt}} = 3$.

As a consequence, the minimization problem is reduced to the examination of two functions:

$$\text{for } x_1 > 0, \qquad f_{d1}(x_1) = 2x_1 + \frac{x_1^2}{3} \tag{1}$$

$$\text{for } x_1 < 0, \qquad f_{d2}(x_1) = 4x_1 + \frac{x_1^2}{3} \tag{2}$$

Using simple differentiation, we find:

– from (1), $x_1^{opt} = -3$ which cannot be true as it contradicts the assumption $x_1 > 0$;

– from (2), $x_1^{opt} = -6$.

Thus, the required solution is: $x_1^{opt} = -6$, $x_2^{opt} = 4$, $x_3^{opt} = 3$.

It should be pointed out that this kind of decomposition procedures is by no means limited to the elimination of discrete variables. If x_2 and x_3 in the above example were continuous, e.g. $x_2 \in [2, 4]$ and $x_3 \in [1, 3]$, then the procedure would be exactly the same, with equally good results.

Example 4.6 (from an unpublished paper by Z. Kotas)

This is a rather complicated problem similar to the problem of optimization of a planetary gear described in [22]. As it turns out, the elimination of redundant constraints and subsequent decomposition make it possible to solve the problem with the considerably reduced computational effort.

Let us minimize

$$f(x_1, x_2, x_3, x_4) = \frac{\pi}{4} x_1 x_2^2 x_3^2 \left(\frac{x_4}{2} + 1 \right)$$

subject to constraints

$$0.6 \leqslant x_1 \leqslant 3.0 \tag{4.8}$$

$$0.25 \leqslant x_2 \leqslant 0.4 \text{ (discrete values spaced by 0.05)} \tag{4.9}$$

$$28 \leqslant x_3 \leqslant 48 \text{ (discrete values spaced by 2)} \tag{4.10}$$

$$x_4 \in \{2, 3, 4\} \tag{4.11}$$

$$10.78 x_1 x_2^2 x_3 x_4 - \sqrt{x_2 x_3} - 2.2 \geqslant 0 \tag{4.12}$$

$$0.21 x_1 x_2^2 x_3^2 x_4 - \sqrt{x_2 x_3} - 2.2 \geqslant 0 \tag{4.13}$$

$$0.0035 x_1 x_2^2 x_3^2 - 1 \geqslant 0 \tag{4.14}$$

$$0.5 x_3 \left(3 \sin \frac{\pi}{x_4} - 2 \right) \geqslant 0 \tag{4.15}$$

The product appearing in the objective function, i.e.

$$C = x_1 x_2^2 x_3^2$$

can be seen to appear also in constraint (4.14) which can be transformed into

$$C \geqslant \frac{10\,000}{35}$$

By comparing constraints (4.12) and (4.13) we note their common component

$$-E = -\sqrt{x_2 x_3} - 2.2$$

It follows from (4.9) and (4.10) that the greatest possible value of E is 6.58.

According to (4.8)–(4.11), the smallest possible value of the first component in (4.13) is 12.35; constraint (4.13) is thus redundant. Similarly, the smallest possible value of the first component in (4.12) is 22.638; this contraint is thus redundant.

Turning again our attention to the objective function we note that it attains its minimum, provided that C and x_4 assume their smallest values. It can thus be concluded that at the point of minimum

$$x_1 x_2^2 x_3^2 = \frac{10\,000}{35}$$

or, after a trivial rearrangement

$$x_1 = \frac{10\,000}{35 x_2^2 x_3^2}$$

Upon examination of constraint (4.15), we can now establish the smallest possible value of x_4. As (4.15) is satisfied for all x_4 values as defined by (4.11), the optimal value of x_4 is 2.

Substituting now $x_4 = 2$ and applying all the above conclusions, we can reduce the original problem to a simple form

$$0.6 \leqslant x_1 \leqslant 3.0$$

$$0.25 \leqslant x_2 \leqslant 0.4 \quad \text{(spaced by 0.05)}$$

$$0.28 \leqslant x_3 \leqslant 48 \quad \text{(spaced by 2)}$$

$$x_1 = \frac{10\,000}{35 x_2^2 x_3^2}$$

$$f_d(x_1, x_2, x_3) = \frac{\pi}{2} x_1 x_2^2 x_3^2 \to \min_{x_1, x_2, x_3}$$

Calculating x_1 values at all feasible values of x_2 and x_3, we can easily establish that the objective function attains its minimum at 15 points (Table 4.1). The smallest function value is $\frac{\pi}{7}$ 1000.

Table 4.1

Values of variable x_1 at points which represent solutions to Example 4.6

x_3	$x_2 = 0.25$	0.30	0.35	0.40
28			2.9749	2.2777
30			2.5910	1.9840
32			2.2710	
34		2.7461	2.0176	
36		2.4490		
38		2.1980		
40	2.8571	1.9841		
42	2.5915			
44	2.3612			
46	2.1604			
48	1.9840			

CHAPTER 5

Methods of finding the extremum of a function

5.1 SURVEY OF EXISTING METHODS

A variety of methods have been devised for finding the extremum of a function, since different combinations of the properties of decision variables, constraints and the objective function result in problems which require different approaches for their solution. A properly selected method may substantially improve both the cost of finding the solution and the accuracy of the results. An unwise selection may cause, in an extreme case, incorrect results to be obtained, for example a local extremum could be found instead of the global one.

Many optimization methods have been developed and refined in the recent years. This has produced a great deal of confusion in the extensive literature [6, 15, 16, 26, 48, 60, 61, 90] now available. In addition to the inconsistent terminology and classification principles, one can find contradictory estimates of the efficiency of certain methods (see [1] and [13]) In this book an attempt is made to resolve such problems by classifying the methods on the basis of their most important features. For each method discussed, a presentation is made of its advantages and disadvantages which require consideration taking into account the nature of the problem to be solved.

There are two basically different approaches to the solution of problems of finding the extremum of a function:

(1) Derivation of a set of equations presenting the necessary conditions for the existence of an extremum and subsequently solving these equations in order to find the optimal values of the variables. This is possible only if continuous first derivatives of the objective function exist. In problems including constraints, it is also necessary that the functions appearing in the constraints be continuous.

(2) Performing an orderly series of numerical experiments in which

values of the objective function are computed (depending on the method used, also the values of its derivatives, gradient, etc.) and finally identifying the optimal values of the variables on the basis of the results obtained. Note that some of the numerous methods in which this approach is used are applicable to functions whose first derivatives are not continuous or are even discontinuous.

Methods which employ the first approach will be classed as indirect, while the second approach is characteristic of the so-called direct methods. The types of problems to be solved can be classified into three groups, namely:

– those requiring the unconstrained extremum (i.e. extremum of a function without constraints),

– those requiring an extremum subject to equality constraints,

– mathematical programming problems (i.e. requiring the extremum of a function with equality and inequality constraints).

The indirect methods are sometimes called analytical methods, but, this term is misleading, since in practical applications the set of equations describing the extremum can rarely be solved analytically. Indirect methods are unlikely to be applicable to problems involving many variables and constraints where the initial analytical transformations leading to the formulation of a set of equations can be tedious and time-consuming. Should this set of equations require numerical methods for its solution, the cumulative workload needed to find the extremum becomes substantially greater than when direct methods are applied. It is an advantage of indirect methods that they allow the global extremum and all points of local extremum to be determined, because the conditions for the existence of the extremum are valid for all feasible solutions.

The direct methods differ with regard to the rules governing the arrangement of the series of numerical experiments. Some methods are well suited to solving both unconstrained and constrained problems; others can be used only in solving problems of a particular type. The direct methods can be grouped into those involving:

– search along the coordinates,
– random search,
– systematic methods,
– combined methods,
– dynamic programming.

Most direct methods locate the extremum without specifying whether it is global or local. In order to determine its nature, additional measures are required such as analytical investigations of the properties of the function or a suitable arrangement of calculations. The selection of a particular direct method for solving a given problem is preceded by an investigation of its mathematical features, i.e. the properties of the decision space, the set of constraints and the objective function, the possibilities of existence of local extreme points, and requirements imposed on the accuracy of results, etc. There is no universally effective and accepted direct method of finding the extremum of a function. The different rules of arrangement of numerical experiments adopted in different methods suggest that, in each of them, the relationships between the amount of computational effort and number of variables, number of constraints and number of arithmetic operations needed to calculate a single function value, etc., are different.

Once a method is selected, it is advisable to consider applying various approaches (see Chapter 4) to the formulation of the problem in order to facilitate an efficient use of the method. Experience shows that such a preliminary treatment may be quite significant. An example (due to J. Oderfeld) is given in [62] of optimization of the telescopic outrigger of a building crane. Suitably modified, this problem was solved at a relatively low cost. Had it been solved, however, in its original form as derived from a design problem, the cost of computer time would have been several hundred times greater.

5.2 INDIRECT METHODS

5.2.1 Unconstrained extremum

As mentioned above, indirect methods of finding the extremum can be applied only to a continuous objective function whose first derivatives are also continuous. At an unconstrained extremum of such a function, the following set of n equations, in the unknowns $x_1, x_2, \ldots, x_n$, must be satisfied

$$\frac{\partial f(\boldsymbol{x})}{\partial x_i} = 0 \qquad i = 1, 2, \ldots, n \tag{5.1}$$

It is possible that equations (5.1) have more than one solution. The points satisfying that set of equations are called *stationary points of function f*. Those points can be points of local maximum, points of local minimum or saddle points. In order to identify the point of global extremum, i.e. the required solution, it is necessary to determine the function values at all stationary points. If the second derivatives of the function are available, the nature of stationary points can be determined using the quadratic form of the matrix

$$\left[\frac{\partial^2 f(\mathbf{x})}{\partial x_i \, \partial x_j}\right] \qquad \begin{array}{l} i = 1, 2, \ldots, n \\ j = 1, 2, \ldots, n \end{array} \tag{5.2}$$

Generally, a quadratic form of an $n \times n$ symmetric matrix $[\boldsymbol{a}_{ij}]$ is defined by the expression:

$$\mathbf{F} = \sum_{i=1}^{n} \sum_{j=1}^{n} a_{ij} \xi_i \xi_j \tag{5.3}$$

where $\xi_1, \xi_2, \ldots, \xi_n$ are arbitrary real variables. If, for any values of variables ξ_i (provided at least one of them does not equal zero), the value of **F** is always positive, the quadratic form is said to be positive-definite. Similarly, if the value of **F** is always negative, the quadratic form is negative-definite. If the sign of **F** varies with the values of variables ξ_i, the quadratic form is indefinite.

Returning now to matrix (5.2), its quadratic form determined for the values of the second derivatives of the function at a stationary point can be:

– positive-definite, indicating that it is a minimum,
– negative-definite, indicating that it is a maximum,
– indefinite, indicating that it is a saddle point.

The use of second derivatives is almost self-explanatory because they determine function properties at points where first derivatives vanish. It is left for the reader to check the details of the identification of the extremum type, by analyzing Taylor series expansion of the function, with the first derivatives equal to zero.

Example 5.1

Find the stationary points of

$$f(x_1, x_2) = x_1^3 - 4x_1^2 + 2x_1 x_2 - x_2^2$$

and determine their nature.

The first and second partial derivatives are:

$$\frac{\partial f}{\partial x_1} = 3x_1^2 - 8x_1 + 2x_2$$

$$\frac{\partial f}{\partial x_2} = 2x_1 - 2x_2$$

$$\frac{\partial^2 f}{\partial x_1^2} = 6x_1 - 8$$

$$\frac{\partial^2 f}{\partial x_2^2} = -2 \qquad \frac{\partial^2 f}{\partial x_1 \, \partial x_2} = 2$$

By solving the set of equations corresponding to (5.1), that is

$$3x_1^2 - 8x_1 + 2x_2 = 0$$
$$2x_1 - 2x_2 = 0$$

we find that two stationary points exist with the coordinates $x_1 = 0$, $x_2 = 0$ and $x_1 = 2$, $x_2 = 2$. At the first point, the function value equals zero and the matrix of the second partial derivatives is:

$$\begin{bmatrix} -8 & 2 \\ 2 & -2 \end{bmatrix}$$

The quadratic form of this matrix is given by:

$$\mathbf{F} = -8\xi_1^2 + 4\xi_1\xi_2 - 2\xi_2^2$$

It can be proved that if $\xi_1 \neq 0$ or $\xi_2 \neq 0$, the value of $\mathbf{F}$ is always negative, i.e. the quadratic form is negative-definite. Thus, the stationary point $x_1 = 0$, $x_2 = 0$ is a maximum. Indeed, small changes in $\boldsymbol{x}$ will always decrease the value of the function: while $f(0, 0) = 0$, we have $f(1, 0) = -3$, $f(0, 1) = -1$, $f(1, 1) = -2$ etc. It is a local maximum, though, as it can be readily observed that, for constant x_2, the function value tends to increase without limit as x_1 increases.

At the second stationary point, the function value is -4 and the matrix of the second partial derivatives is

$$\begin{bmatrix} 4 & 2 \\ 2 & -2 \end{bmatrix}$$

The quadratic form, expressed by

$$\mathbf{F} = 4\xi_1^2 + 4\xi_1\xi_2 - 2\xi_2^2$$

is indefinite, so the stationary point $x_1 = 2$, $x_2 = 2$ is a saddle point. Indeed, some changes in $\boldsymbol{x}$ will increase the function value – for example, if x_1 is changed to 1, then $f(1, 2) = -3$, when others will decrease it – for example, if x_2 is changed to 1, then $f(2, 1) = -5$.

5.2.2 Equality constrained extremum

This problem can be solved using the Lagrange multiplier method. The objective function, $f(\boldsymbol{x})$, and the set of constraints

$$g_i(\boldsymbol{x}) = 0, \qquad i = 1, 2, \ldots, m \tag{5.4}$$

where m is less than the dimension, n, of the system, are combined into an augmented objective function:

$$f_a(\boldsymbol{x}, \lambda_1, \lambda_2, \ldots, \lambda_m) = f(\boldsymbol{x}) - \lambda_1 g_1(\boldsymbol{x}) - \lambda_2 g_2(\boldsymbol{x}) \ldots - \lambda_m g_m(\boldsymbol{x}) \tag{5.5}$$

Here λ_i are Lagrange multipliers, independent of $\boldsymbol{x}$, to which certain values will be assigned in the course of calculations. The conditional stationary point of function f exists at the point which is an unconstrained stationary point of the augmented function f_a. It must, therefore, satisfy the set of equations:

$$\begin{cases} \dfrac{\partial f_a}{\partial x_i} = 0 & i = 1, 2, \ldots, n \\ \dfrac{\partial f_a}{\partial \lambda_j} = 0 & j = 1, 2, \ldots, m \end{cases} \tag{5.6}$$

which can be written in the form:

$$\begin{cases} \dfrac{\partial f(\boldsymbol{x})}{\partial x_i} - \displaystyle\sum_{j=1}^{m} \lambda_j \frac{\partial g_j(\boldsymbol{x})}{\partial x_i} = 0 & i = 1, 2, \ldots, n \\ g_j(\boldsymbol{x}) = 0 & j = 1, 2, \ldots, m \end{cases} \tag{5.7}$$

In general, the augmented function may have several stationary points. Determination of the extremum may thus involve calculating and comparing several function values, as in the case of an unconstrained extremum.

Example 5.2

Find the minimum of the function:

$$f(x_1, x_2) = -x_1^2 + x_2$$

subject to the constraint

$$x_1^2 + x_2^2 - 1 = 0$$

The augmented function is:

$$f_a(x_1, x_2, \lambda) = -x_1^2 + x_2 - \lambda(x_1^2 + x_2^2 - 1)$$

The necessary conditions for the minimum to exist are:

$$\begin{cases} -2x_1 - \lambda 2x_1 = 0 \\ 1 - \lambda 2x_2 = 0 \\ x_1^2 + x_2^2 - 1 = 0 \end{cases}$$

This set of equations can be easily solved because it is immediately seen that if $x_1 = 0$, then x_2 is obtained from the third equation and λ from the second. If $x_1 \neq 0$, then the first and the second equation can be used to eliminate λ and x_1, making it again possible to determine x_2 from the third equation. As this equation is quadratic, two x_2 values are obtained in each case, so the set of equations has four solutions:

$$(1) \quad x_1 = 0 \qquad x_2 = 1 \qquad \lambda = \frac{1}{2}$$

$$(2) \quad x_1 = 0 \qquad x_2 = -1 \qquad \lambda = -\frac{1}{2}$$

$$(3) \quad x_1 = \frac{\sqrt{3}}{2} \qquad x_2 = -\frac{1}{2} \qquad \lambda = -1$$

$$(4) \quad x_1 = -\frac{\sqrt{3}}{2} \qquad x_2 = -\frac{1}{2} \qquad \lambda = -1$$

The corresponding function values are: $+1$, -1, $-\frac{5}{4}$, $-\frac{5}{4}$. Thus, the function has two equivalent conditional minima at the points $\left(\frac{\sqrt{3}}{2}, -\frac{1}{2}\right)$ and $\left(-\frac{\sqrt{3}}{2}, -\frac{1}{2}\right)$. This can be seen in Fig. 5.1 which also presents the geometrical meaning of Lagrange equations (5.7), namely that the objective function contour touches the constraint at the solution.

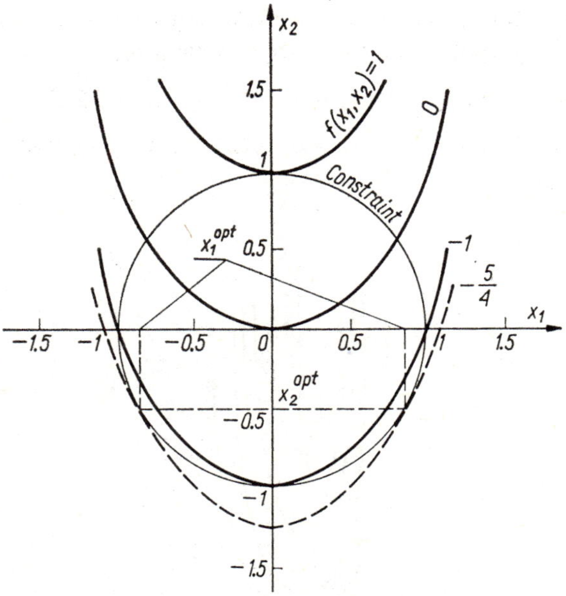

Fig. 5.1 Illustration of Example 5.2.

Example 5.3

Find the minimum of the function:

$$f(x_1, x_2) = x_1^2 - \ln(x_2 - 2)$$

subject to the constraints:

$$x_2 \geqslant 2 \qquad x_1+x_2 = 6$$

The inequality constraint may be temporarily neglected if a new non-negative variable $x_n = x_2-2$ is introduced. The function, the equality constraint and the resulting augmented function assume then the form:

$$f(x_1, x_n) = x_1^2-\ln x_n$$

$$x_1+x_n-4 = 0$$

$$f_a(x_1, x_n, \lambda) = x_1^2-\ln x_n-\lambda(x_1+x_n-4)$$

The conditions necessary for the minimum to exist are:

$$\begin{cases} 2x_1-\lambda = 0 \\ -\dfrac{1}{x_n}-\lambda = 0 \\ x_1+x_n-4 = 0 \end{cases}$$

This set of equations can be solved using the first and the second equation to eliminate λ and x_1. From the quadratic equation thus obtained, x_n can be calculated. Consequently, there are two solutions

(1) $x_n = -0.121$ $\quad x_1 = 4.121$ $\quad \lambda = 8.242$

(2) $x_n = 4.121$ $\quad x_1 = -0.121$ $\quad \lambda = -0.242$

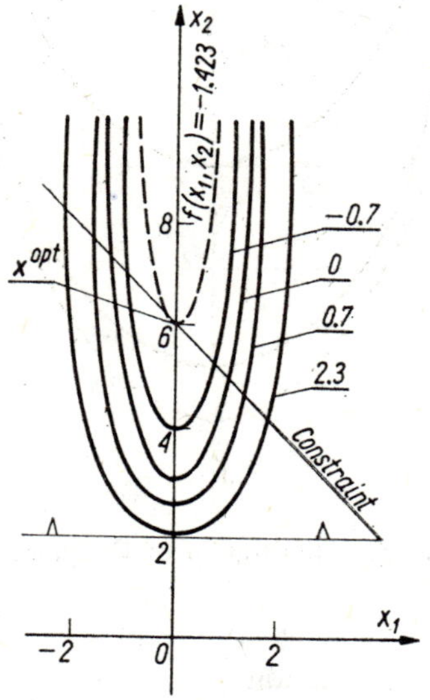

Fig. 5.2 Illustration of Example 5.3.

The first solution is invalid as the value of x_n is negative which is inconsistent with the assumption. The second solution, $x_n = 4.121$, corresponds to $x_2 = 6.121$ so it can be concluded that there is a stationary point at $(-0.121, 6.121)$. The function value at this point is -1.423 and it can be easily proved that it is a conditional minimum, as can also be seen in Fig. 5.2.

Note that it is possible to obtain the same result by substituting (from the constraint)

$$x_n = 4 - x_1$$

The problem thus reduces to finding the minimum of the function of one variable

$$f(x_1) = x_1^2 - \ln(4 - x_1)$$

and then determining x_2 from the equality constraint.

5.2.3 Inequality constrained extremum

Inequality constrained problems are usually solved by one of the many methods of mathematical programming. The necessary conditions for the existence of the minimum, known as Kuhn–Tucker conditions, can be used for solving such problems but they are seldom useful for effective determination of the solutions. One possible exception is the class of problems featuring a monotonic objective function. Such problems can be solved using the so-called monotonicity tables based on the Kuhn–Tucker conditions [87]. In many other cases of applicability of these conditions (when the objective function and the functions appearing in the constraints are continuously differentiable), efficient direct methods are available. The computational effort needed to find the solution sharply increases with the number of variables and constraints but the Kuhn–Tucker conditions are still of great theoretical importance, particularly in the investigation of the general properties of certain classes of optimization problems. They are also useful in the construction of algorithms and in the analysis of their efficiency.

For a problem of the form

$$\min f(\boldsymbol{x})$$

subject to (5.8)

$$g_j(\boldsymbol{x}) \leqslant 0 \qquad j = 1, 2, \ldots, m$$

the Kuhn–Tucker conditions may be stated as follows

$$\frac{\partial f(\boldsymbol{x})}{\partial x_i} + \sum_{j=1}^{m} \lambda_j \frac{\partial g_j(\boldsymbol{x})}{\partial x_i} = 0 \qquad i = 1, 2, \ldots, n$$

$$g_j(\boldsymbol{x}) \leqslant 0 \qquad (5.9)$$
$$\lambda_j \geqslant 0 \qquad j = 1, 2, \ldots, m$$
$$\lambda_j g_j(\boldsymbol{x}) = 0$$

where λ_j are multipliers which assume the value zero when their associated constraints are inactive, that is satisfied as strict inequalities; otherwise their values are positive. The Kuhn–Tucker conditions are meaningful only if, in addition to the requirements of continuity and differentiability of the functions involved, the requirement of the so-called regularity of the constraint system [15] is also satisfied. The application of these conditions consists of determining (by analysing the equations and inequalities (5.9) and determining acceptable values of the multipliers) those constraints which are active i.e. against which the extremum lies. Subsequently, the points at which the extremum can be found are identified. As some of these points can be points of local extremum, the corresponding values of the objective function should be determined and compared.

The examples given below illustrate the application of the Kuhn–Tucker conditions only indirectly, since instead of relations (5.9) use is made of the approach which, in the general case, leads to the formulation of the conditions. At the same time it is demonstrated how the complexity increases with the number of constraints.

Example 5.4

Find the minimum of the function

$$f(x_1, x_2) = (x_1-1)^2+(x_2-2)^2+1$$

subject to the constraint

$$x_1+x_2-2 \leqslant 0$$

The inequality constraint can be converted to an equality constraint by introducing a non-negative slack variable x_s:

$$x_1+x_2-2+x_s = 0$$

Combining the objective function and the converted constraint we get an augmented objective function:

$$f_a(x_1, x_2, x_s, \lambda) = (x_1-1)^2+(x_2-2)^2+1-\lambda(x_1+x_2-2+x_s)$$

If the solution lies inside the feasible region (case I) then $x_s \neq 0$ and the derivatives of f_a with respect to x_1, x_2 and x_s must equal zero. On the other hand, if the solution lies on the boundary (case II), then $x_s = 0$ and the derivatives with respect to x_1, x_2 and λ vanish. The two relevant sets of equations should be considered separately:

(I) $2(x_1-1)-\lambda = 0$ (II) $2(x_1-1)-\lambda = 0$

$2(x_2-2)-\lambda = 0$ $2(x_2-2) = 0$

$\lambda = 0$ $x_1+x_2-2 = 0$

Note that the set (I) is obtained from the Kuhn–Tucker conditions by setting $\lambda = 0$ (constraint inactive), while (II) corresponds to the alternative situation. The solution of the set (I) yields $x_1 = 1$, $x_2 = 2$, $\lambda = 0$, which cannot be an extremum point as the constraint is not satisfied there. On the other hand, the solution of set (II), i.e. $x_1 = 1/2$, $x_2 = 3/2$, $\lambda = -1$, is feasible and so the minimum lies on the boundary of the feasible region and the lowest value of the function is 3/2 (Fig. 5.3).

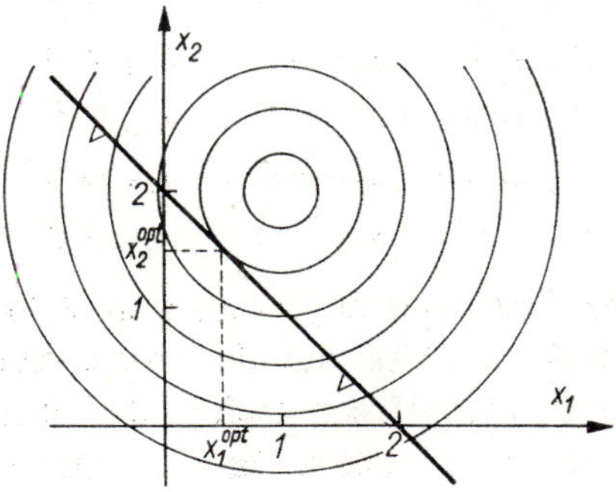

Fig. 5.3 Illustration of Example 5.4.

Example 5.5

Find the minimum of the function

$$f(x_1, x_2) = 10(x_1-3.5)^2+20(x_2-4)^2$$

subject to the constraints

$$x_1 \leqslant 6-x_2$$

$$x_2 \geqslant x_1-1$$

$$x_1 \geqslant 0$$

The constraints can be converted to equality constraints by using three non-negative slack variables x_{s1}, x_{s2}, x_{s3}:

$$x_1+x_2-6+x_{s1} = 0$$

$$x_1-x_2-1+x_{s2} = 0$$

$$-x_1+x_{s3} = 0$$

Thus the augmented objective function assumes the form:

$$f_a(x_1, x_2, x_{s1}, x_{s2}, x_{s3}, \lambda_1, \lambda_2, \lambda_3) = 10(x_1-3.5)^2 +$$
$$+20(x_2-4)^2-\lambda_1(x_1+x_2-6+x_{s1})-\lambda_2(x_1-x_2-1+x_{s2})-\lambda_3(-x_1+x_{s3})$$

Consider first the situation where the solution lies inside the feasible region, in which case all the slack variables will differ from zero and the derivatives of f_a with respect to $x_1, x_2, x_{s1}, x_{s2},$ and x_{s3} will vanish. Alternatively, if the solution lies on the boundary of the region, then the following different possibilities arise

(II)	$x_{s1} = 0$	$x_{s2} \neq 0$	$x_{s3} \neq 0$
(III)	$x_{s1} = 0$	$x_{s2} = 0$	$x_{s3} \neq 0$
(IV)	$x_{s1} = 0$	$x_{s2} \neq 0$	$x_{s3} = 0$
(V)	$x_{s1} \neq 0$	$x_{s2} = 0$	$x_{s3} \neq 0$
(VI)	$x_{s1} \neq 0$	$x_{s2} = 0$	$x_{s3} = 0$
(VII)	$x_{s1} \neq 0$	$x_{s2} \neq 0$	$x_{s3} = 0$
(VIII)	$x_{s1} = 0$	$x_{s2} = 0$	$x_{s3} = 0$

but in each case the derivatives of f_a with respect to $x_1, x_2,$ the appropriate (non-zero) slack variables and the corresponding multipliers are expected to vanish. Cases I–VIII can each be described by a set of several equations involving several variables. Once the individual solutions have been found, it is necessary to check whether they satisfy the constraints given in the original formulation of the problem. Only in this manner can the case corresponding to the required solution be identified.

As can be easily demonstrated, the solution of the above problem with the aid of the Kuhn–Tucker conditions would require less work, but still it would require the construction and analysis of five equations and six inequalities, corresponding to set (5.9), which is quite tedious.

Although applicable, there are features of the above problem which remove the necessity for the general approach described above. As this is a two-dimensional problem, the constraints can easily be represented graphically. Owing to the simple form of the function, the same can be said of its contours or isolines. From Fig. 5.4 we can conclude immediately that the solution of the problem corresponds to case (II), i.e. it can be described by the set of equations (neglecting the equations used for determination of x_{s2} and x_{s3}:)

$$\begin{cases} 20(x_1-3.5)-\lambda_1 = 0 \\ 40(x_2-4)-\lambda_1 = 0 \\ x_1+x_2-6 = 0 \end{cases}$$

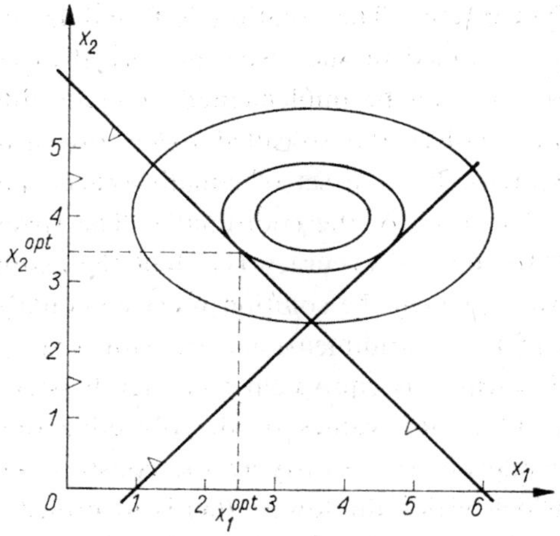

Fig. 5.4 Illustration of Example 5.5.

Solution of this set of equations yields $x_1 = 2.5$, $x_2 = 3.5$ and $\lambda_1 = -20$ at which point the function takes its minimum value of 15.

5.3 DIRECT METHODS

Before studying this section it is recommended to consult Appendices 2 and 3, containing a treatment of vector and matrix notation and the use of iterative formulae. For the sake of brevity, it is assumed throughout Section 5.3 that the optimization problem involves minimization of the objective function.

In general, the algorithms based on individual methods are presented below in sufficient detail to demonstrate the most important features of a particular group of methods. In order to prepare computer programs implementing individual algorithms and methods, it may be necessary to consult the literature. However, it should be stressed that it is not easy to produce robust and efficient procedures for finding an extremum. One should avoid writing new programs if the required methods are available in the program library of the computing centre. It is also advisable to make use of published listings of optimization programs in ALGOL or FORTRAN [12, 40, 43, 71].

5.3.1 Search in the direction of coordinates

A common feature of the methods considered here is the partition of a series of numerical experiments into sections in which the search is performed with respect to one variable at a time, the values of remaining variables being temporarily constant. The solution is found by direct comparison of function values obtained in successive numerical experiments.

The above scheme can be implemented in many different ways. The simplest technique involves the so-called *exhaustive search* (also known as the factorial search). Here, a set of equally spaced points in a certain interval is assigned to each of the coordinates. The spacing of the points should equal half the desired accuracy (i.e. half the assumed width of the uncertainty interval in which the solution must be contained). In this way a grid is defined in the multidimensional decision space.

The exhaustive search is equivalent to a step-by-step *tabulation of the objective function.* Once the values of the objective function have been calculated for all relevant (i.e. satisfying the constraints) grid points, the grid value of $\boldsymbol{x}$ which yields the lowest $f(\boldsymbol{x})$ is identified as the minimum. This is shown for the one-dimensional case in Fig. 5.5.

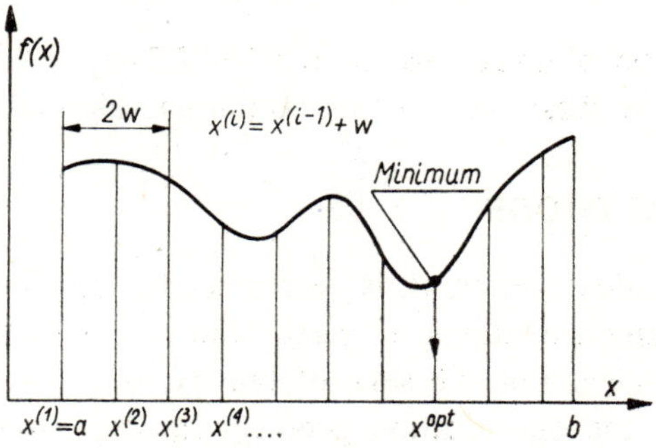

Fig. 5.5 Exhaustive search in the interval $[a, b]$; the accuracy of minimization is $2w$ (or $\pm w$).

The described search technique can be easily applied to unconstrained problems or problems with inequality constraints. When equality constraints occur, a special approach is necessary. Consider the problem of the form

$$\min_{\boldsymbol{x}} f(\boldsymbol{x}) \qquad \boldsymbol{x} = (x_1, x_2, \ldots, x_n)$$

$$\text{subject to } g_i(\boldsymbol{x}) = 0 \qquad i = 1, 2, \ldots, p; \quad p < n \tag{5.10}$$

$$h_j(\boldsymbol{x}) \leqslant 0 \qquad j = 1, 2, \ldots, q$$

If the equality constraints make it possible to eliminate p variables, then after appropriate substitutions the simplified problem is obtained:

$$\min_{\boldsymbol{y}} t(\boldsymbol{y}) \qquad \boldsymbol{y} = (y_1, y_2, \ldots, y_{n-p})$$

$$\text{subject to } u_j(\boldsymbol{y}) \leqslant 0 \qquad j = 1, 2, \ldots, q \tag{5.11}$$

where: y are the variables left after the elimination, t and u_j are the functions f and h_j suitably reduced by elimination of some variables. The simplified problem can be solved by exhaustive search in the $(n-p)$-dimensional space. On the other hand, if the equality constraints for some reason do not allow analytical substitutions for variable elimination, then only $(n-p)$ variables can be regarded as independent in the search. The values of the remaining p variables must be determined from the set of equality constraints. The application of this principle is shown as a logic diagram in Fig. 5.6.

The exhaustive search offers such important advantages as simplicity of the search algorithm and the robustness against awkward mathematical

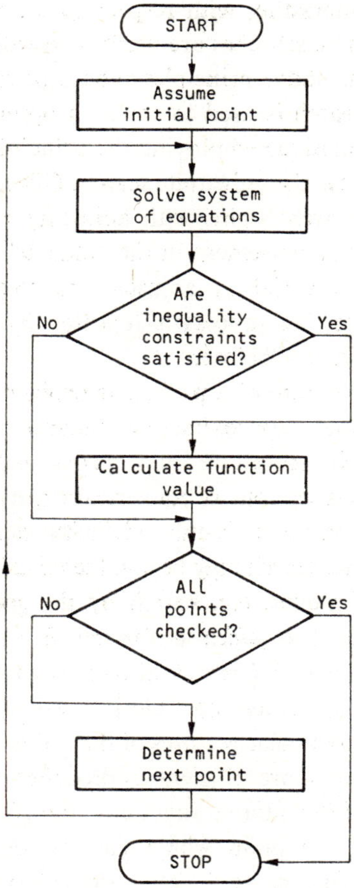

Fig. 5.6 Logic diagram of the exhaustive search for problems with equality and inequality constraints.

properties of the objective function and constraints. Owing to the fact that the method involves searching the whole region of feasible solutions, it has the ability to recognize approximately the global properties of the objective function, including the multiple local extremum points. The main drawback of the method is, however, the fact that the computational effort increases exponentially with the number of variables. Moreover, the results may involve a significant error when the intervals of variable values and the search accuracy are not evaluated properly.

The disadvantages of the exhaustive search can be partly offset by multiple application of the method accompanied by reduction of the width of the intervals of variable values. Each time the search is concluded, the results are used to identify, in the feasible region, those parts which can be discarded as less interesting with respect to the value of the objective function. The consecutive searches are performed with increasing accuracy. As a matter of fact, the above principles should be routinely adopted whenever the exhaustive search is used to solve an optimization problem.

Another search technique employing the principle of proceeding in the direction of coordinates is the univariate search. Given certain initial settings of the variables, the search begins with adjusting x_1 by one-dimensional search until no further improvement in the value of the objective function is obtained; then another variable is adjusted, and so on. After each variable has been adjusted once, the process is repeated by returning to the first variable and proceeding as before.

Depending on the nature of a particular problem (whether $f(\boldsymbol{x})$ is continuous or discontinuous, approximately quadratic or not) a particular one-dimensional search procedure may prove more appropriate than others for the individual sections of numerical experiments. If the objective function in the region considered is unimodal with respect to all the variables, then the one-dimensional search may be conducted using one of the so-called interval reduction techniques, e.g. search by the golden section or Fibonacci search [40, 49, 60]. The *golden section search* has a somewhat simpler algorithm, while the *Fibonacci search* allows us to determine, after completing n iterations, the smallest possible interval of uncertainty in which the optimum lies. The particular features of the objective function may allow the application of even more efficient approaches to interval reduction, e.g. for continuously differentiable functions, the *Newton–Raphson method* can be used to locate the point where the first derivative vanishes (see Appendix 3). The basic idea underlying the interval reduction methods is trapping of the minimum in successively decreasing, nested intervals $[a_i, b_i]$. This is done by evaluating f at two interior points l_i and r_i (left and right one) of the current interval; then, for a unimodal function, it is always possible to identify the subinterval – either $[a_i, l_i]$ or $[r_i, b_i]$ – in which the minimum cannot possibly lie (Fig. 5.7). Thus, either $[a_i, r_i]$ or $[l_i, b_i]$ is assumed as the new smaller interval $[a_{i+1}, b_{i+1}]$. The interior points are determined from the formulae:

$$l_i = b_i - t_i(b_i - a_i) \tag{5.12}$$

$$r_i = a_i + t_i(b_i - a_i) \tag{5.13}$$

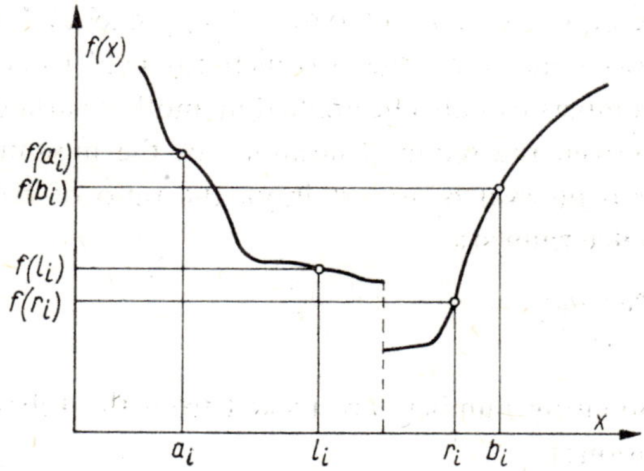

Fig. 5.7 Detection of a subinterval in which the minimum cannot lie: the left-hand side of $[l_i, r_i]$ is eliminated since for a unimodal function the values $f(a_i)$, $f(l_i)$ and $f(r_i)$ indicate that the function increases in that direction.

In the search by the golden section method it is assumed for all values of i that

$$t_i = \frac{\sqrt{5}-1}{2} = 0.618$$

while for the Fibonacci search

$$t_i = \frac{F_{n-i-1}}{F_{n-i}} \qquad i = 0, 1, 2, \ldots, n-3 \tag{5.14}$$

where F_j are Fibonacci numbers ($F_0 = F_1 = 1$, $F_2 = 2$, $F_3 = 3, \ldots, F_j = F_{j-1} + F_{j-2}, \ldots$). In the $(i+1)$th iteration the rules for defining the interval containing the minimum are:

$$\begin{aligned} &\text{if} \quad f(l_i) > f(r_i), \quad \text{take} \quad a_{i+1} = l_i,\ b_{i+1} = b_i \\ &\text{if} \quad f(l_i) \leqslant f(r_i), \quad \text{take} \quad a_{i+1} = a_i,\ b_{i+1} = r_i \end{aligned} \tag{5.15}$$

The search may be terminated after the prescribed number of iterations has been performed or when the interval length becomes smaller than the desired accuracy. The latter rule is a typically local criterion which may be satisfied in the vicinity of the global minimum or of local ones. Note that after n iterations the length of the interval containing the minimum is reduced in the case of the golden section search to:

$$b_{n-1}-a_{n-1} = 0.618^{n-1}(b_0-a_0) \tag{5.16}$$

and in the case of the Fibonacci search to:

$$b_{n-1}-a_{n-1} = \frac{1}{F_n}(b_0-a_0) \tag{5.17}$$

For large values of n the value of 0.618^{n-1} approaches $1/F_n$, i.e. as the number of iterations increases, the two methods tend to equal efficiency.

Fibonacci numbers can also be applied in another variant of the search algorithm [10]. Given the required accuracy of the minimum location w (width of the final interval of uncertainty), the ratio of reduction of the initial interval is determined:

$$N = \frac{b_0-a_0}{w} \tag{5.18}$$

Next, two neighbouring numbers are picked from the Fibonacci series to satisfy the condition:

$$F_{n-1} < N \leqslant F_n \tag{5.19}$$

The smallest step of the search is then defined as:

$$w_m = \frac{b_0-a_0}{F_{n-1}} \tag{5.20}$$

The initial step of the algorithm yields the first estimate of the minimum:

$$x_1 = a_0+w_m F_{n-2} \tag{5.21}$$

Further estimates are found from the formula:

$$x_{i+1} = x_i \pm w_m F_{n-i-2} \tag{5.22}$$

where: the sign + is used when $f(x_i) < f(x_{i-1})$, otherwise we use the negative.

In Fig. 5.8, an example is shown of a search sequence resulting from this variant of the Fibonacci search. Seven steps (numbered 0 to 6) were sufficient to obtain an accuracy of about 5% of the initial interval size.

If the objective function can be approximated in the vicinity of the minimum by a quadratic or cubic polynomial, then another group of very efficient methods of one-dimensional search may be applied. These are the so-called methods of substitutional functions, exemplified by the method of successive quadratic approximations (Fig. 5.9).

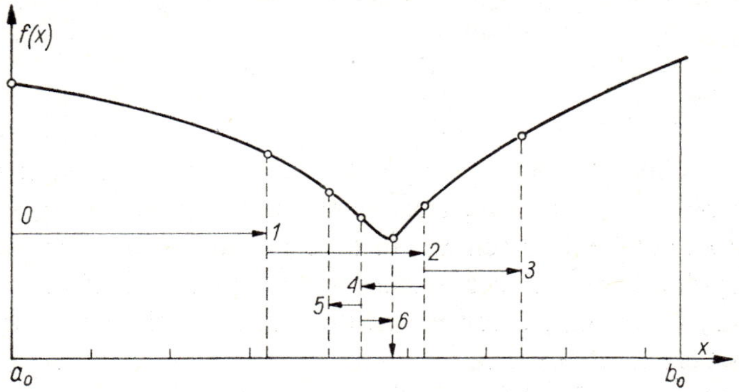

Fig. 5.8 Example of Fibonacci search.

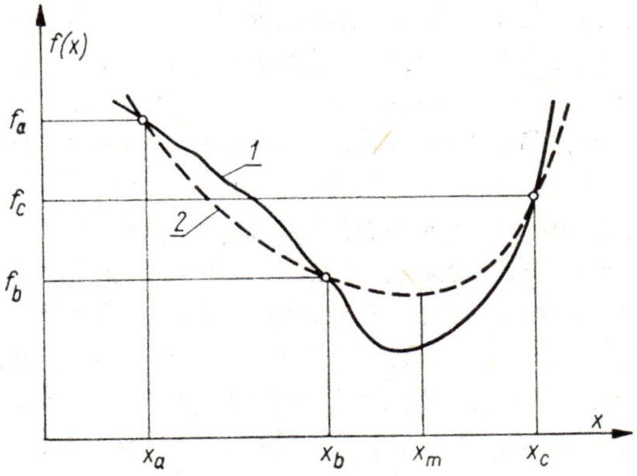

Fig. 5.9 Principle of a single step in the method of successive approximations: 1 – objective function, 2 – interpolation polynomial.

Let the minimum be contained in the interval limited by points x_a and x_c with point x_b given between them, and let the corresponding function values be f_a, f_b, f_c. It is now possible to approximate the function by the Lagrange interpolation polynomial (see Appendix 3)

$$y(x) = f_a \frac{(x-x_b)(x-x_c)}{(x_a-x_b)(x_b-x_c)} + f_b \frac{(x-x_a)(x-x_c)}{(x_b-x_a)(x_b-x_c)} + f_c \frac{(x-x_a)(x-x_b)}{(x_c-x_a)(x_c-x_b)} \quad (5.23)$$

The first derivative of this expression with respect to x, equated to zero, yields the point x_m in which the polynomial attains its minimum:

$$x_m = \frac{1}{2} \frac{(x_b^2-x_c^2)f_a + (x_c^2-x_a^2)f_b + (x_a^2-x_b^2)f_c}{(x_b-x_c)f_a + (x_c-x_a)f_b + (x_a-x_b)f_c} \quad (5.24)$$

In a special case of equally spaced points, i.e. when $x_c - x_b = x_b - x_a = L$, formula (5.24) assumes the form:

$$x_m = x_b - \frac{L}{2} \frac{f_c - f_a}{f_c - 2f_b + f_a} \tag{5.25}$$

The next estimate of the border points of the interval containing the minimum is obtained by rejecting from among the points x_a, x_b, x_c the point which does not border upon x_m. In the new interval, x_m becomes a new internal point. The search is concluded if the displacement of the estimated minimum, i.e. $x_b - x_m$, becomes smaller than the desired accuracy. This, of course, is a local criterion.

It should be pointed out that the sole purpose of the above description is to explain the principle of substitutional methods of one-dimensional search, since other substitutional methods, e.g. Davidon's cubic-convergent search or Powell's quadratic-convergent search [60], are much more efficient. Generally, substitutional methods require more complex calculations than the interval reduction methods, but they ensure faster convergence to the solution. The uncertainty interval decreases proportionally to the second or third power of the number of iterations, while the interval reduction methods are characterized by linear (golden section search) or asymptotically linear (Fibonacci search) convergence. Thus, whenever the properties of the objective function allow, substitutional function methods should be applied instead of interval reduction ones.

Finally, it should be emphasized that the univariate search in a multidimensional space – independently of whether substitutional functions or interval reductions are employed – can be applied only to unconstrained

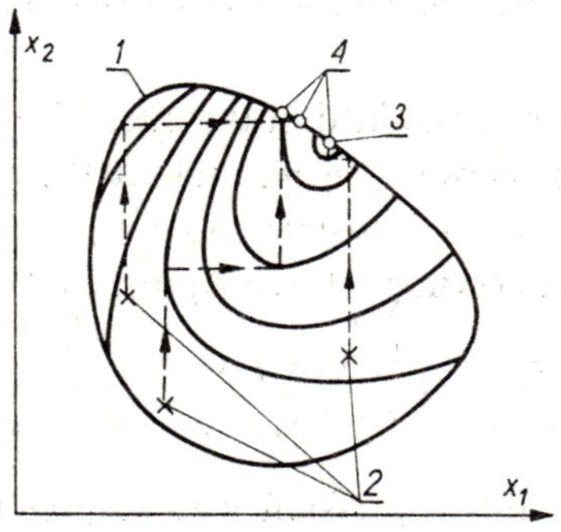

Fig. 5.10 Example of constraint properties which may prevent the univariate search from converging to the extremum: 1 – boundary of the region of feasible solutions, 2 – starting points, 3 – global minimum, 4 – terminal points.

problems or to constrained problems featuring the minimum inside the feasible region. Figure 5.10 shows a constraint configuration which prohibits, for certain initial settings of the variables, the finding of the minimum point at the boundary of the feasible region. The univariate search may also be terminated at a point incorrectly identified as the minimum in the case of function properties such that, at certain points, it is impossible to establish the decrease of the function along any coordinate (Fig. 5.11).

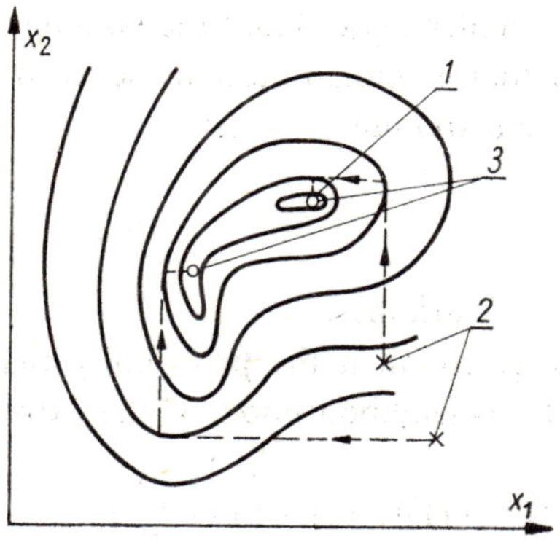

Fig. 5.11 Example of objective function properties which may prevent the univariate search from converging to the extremum: 1 – global minimum, 2 – starting points, 3 – terminal points.

For continuously differentiable functions, it is possible to overcome such difficulties by using various procedures of the so-called *relaxation search* [60] in which the sequence of coordinates being searched may vary depending on the values of the function derivatives.

5.3.2 Random search

It is characteristic of random search methods that, in numerical experiments, the values of the variables are generated with the aid of random numbers. In some methods, the values are selected entirely at random within some region of interest; other methods rely on the selection of points from the immediate vicinity of a current search-point centre. Like in the methods presented in the preceding section, the solution is identified by direct comparison of successively obtained function values.

The simplest random search technique is the so-called *crude Monte Carlo method.* Some assumptions of this method make it similar to the exhaustive search: the values of the variables are also selected from certain chosen intervals (i.e. the method is concerned with the points belonging to a certain region, usually in the form of a parallelepiped, in space X) and the equality constraints require the same kind of manipulation (i.e. separation of the independent variables from those calculated from the set of equations). However, instead of defining a priori the positions of the points within the region, they are selected at random. The number of points should be in proportion to the desired accuracy of location of the extremum.

Let the width of the uncertainty interval of each coordinate be ε (where ε is a number smaller than 1) and the volume of the feasible region be V. The probability that a single randomly selected point lies within the ε-neighbourhood of the extremum is then:

$$p_\varepsilon(1) = \frac{\varepsilon^n}{V} \tag{5.26}$$

where n is the number of variables.

If r points are selected, then the probability that at least one point is contained within the ε-neighbourhood of the extremum is given by:

$$p_\varepsilon(r) = 1-[1-p_\varepsilon(1)]^r = 1-\left(1-\frac{\varepsilon^n}{V}\right)^r \tag{5.27}$$

This expression can be transformed to yield the number r^* of points needed to obtain the desired value p_ε^* of the probability that at least one point lies within the ε-neighbourhood of the extremum:

$$r^* = \frac{\ln(1-p_\varepsilon^*)}{\ln\left(1-\frac{\varepsilon^n}{V}\right)} \approx \frac{V}{\varepsilon^n}\ln\frac{1}{1-p_\varepsilon^*} \tag{5.28}$$

As can be easily seen, the required number of points increases exponentially with the number of variables. This inconvenient feature of the Monte Carlo method may be eliminated by multiple application of the method in a region diminishing in size and with the uncertainty interval narrowing correspondingly (as in the case of the exhaustive search). Such an approach is eventually adopted in a number of random search procedures derived from the Monte Carlo method [62, 63].

In practical situations, the application of formula (5.28) may be somewhat inconvenient because it may be difficult to evaluate volume V when the constraint form is complicated. An intuitive stopping rule that is often used states that the search should be terminated if the growing number of randomly selected points does not produce any new function values that are better than those at points selected earlier.

The crude Monte Carlo method offers similar advantages to the exhaustive search; it has a simple algorithm, is insensitive to the properties of the objective function and reveals an ability to recognize the global properties of the objective function. Its efficiency is close to that of the exhaustive search and decreases when the number of variables becomes large. Being completely unbiased in the selection of search points, however, the Monte Carlo method eliminates the risk of missing an absolute extremum which in the case of exhaustive search may remain undetected due to the arbitrary selection of the grid points.

To help to understand the technique of random selection of points in the domain of interest, imagine a finite series of random numbers $l_1, l_2, l_3, \ldots,$ their values being normalized to [0, 1], i.e. $0 \leqslant l_j \leqslant 1$ for all j. For each variable x_i the border points of the interval of the values of interest are a_i, b_i. The coordinates of the first point selected are as follows:

$$
\begin{aligned}
x_1^{(1)} &= a_1+(b_1-a_1)l_1 \\
x_2^{(1)} &= a_2+(b_2-a_2)l_2 \\
&\vdots \\
x_n^{(1)} &= a_n+(b_n-a_n)l_n
\end{aligned}
\tag{5.29}
$$

The ith coordinate of the mth point can be expressed in general as:

$$
x_i^{(m)} = a_i+(b_i-a_i)l_{n(m-1)+i} \tag{5.30}
$$

In computer applications, the series of random numbers is usually obtained from recurrent formulae known as *random number generators.* Such formulae are available as standard functions or subprograms on most computers. For example, random numbers for relatively short series may be obtained from the formula known as the multiplicative generator:

$$
l_i = cl_{i-1}(\text{mod } 1) \tag{5.31}
$$

where: c is a constant greater than 1, (mod 1) means that the fractional part of the product cl_{i-1} should be taken as the result. To use formula (5.31) properly it is necessary to select appropriate values of c and l_0 since otherwise, in an unfavourable case, parts of the series of numbers obtained may turn out to be geometric sequences.

The flowchart shown in Fig. 5.12 shows the principle of the crude Monte Carlo method. The possibility of manual intervention in the course of the calculation, particularly after the completion of a series of numerical experiments and the evaluation of results, is of great significance; the search can be terminated or another series started with possible modification of the intervals with the individual variables.

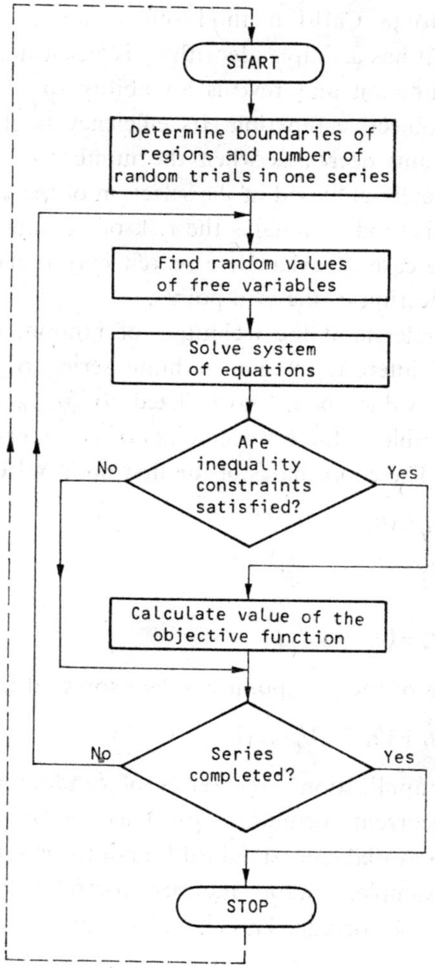

Fig. 5.12 Logic diagram of the crude Monte Carlo method.

The very simple rule governing the numerical experiments in the crude Monte Carlo method makes it vulnerable to certain properties of the constraint set. The risk of reduced efficiency of the method arises from the possibility that an unfortunately determined parallelepiped in space X may turn out to be substantially larger than the feasible region, as shown in Fig. 5.13. Then, most of the points selected at random violate the constraints, which means that computational effort is wasted.

For such problems another group of random search methods is available. These methods incorporate certain safeguards against departure from the feasible region. A typical example is the *random creep method*

[7, 44]. Assuming that we consider only the problems with inequality constraints, i.e. of the same form as (5.11), the method is based on the iterative formula:

$$\boldsymbol{x}^{(i+1)} = \boldsymbol{x}^{(i)} + \Delta\boldsymbol{x}^{(i+1)} \tag{5.32}$$

The displacement vector $\Delta\boldsymbol{x}$ in the above expression is determined depending on whether or not the displaced point remains within the feasible region as the function value decreases. The governing rule for the displacement vector can be formulated as ($j = 1, 2, \ldots, m$)

$$\Delta\boldsymbol{x}^{(i+1)} = \begin{cases} \Delta\boldsymbol{x}^{(i)} & \text{if } f(\boldsymbol{x}^{(i+1)}) \leqslant f(\boldsymbol{x}^{(i)}) \text{ and } h_j(\boldsymbol{x}^{(i+1)}) \leqslant 0 \\ -\Delta\boldsymbol{x}^{(i)} + \boldsymbol{l}\delta_i & \text{if } f(\boldsymbol{x}^{(i+1)}) > f(\boldsymbol{x}^{(i)}) \text{ or } h_j(\boldsymbol{x}^{(i+1)}) > 0 \end{cases} \tag{5.33}$$

where: $\boldsymbol{l}$ is the normalized (i.e. of unit length) random vector, and δ_i is the step length.

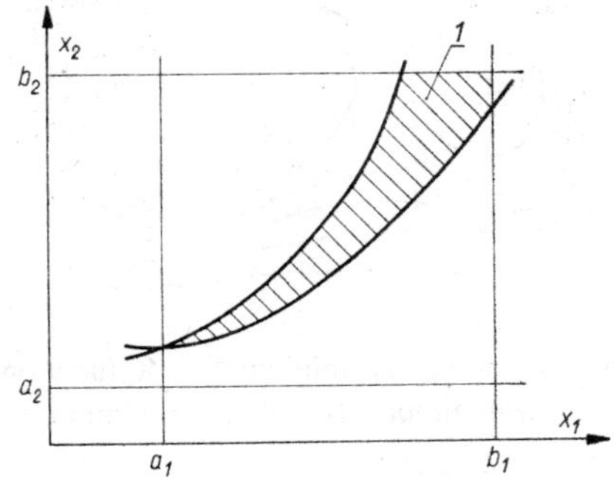

Fig. 5.13 Unfavourable configuration of constraints decreasing the efficiency of the crude Monte Carlo method: 1 – region of feasible solutions.

The step length assumed at the beginning of the search is successively reduced every time a given number of unsuccessful displacement trials have been attempted. The search may be terminated if the length of the displacement vector becomes smaller than the desired accuracy. Using such a stopping rule, the random creep method may – contrary to the crude Monte Carlo algorithm – converge to a local instead of the global extremum. In order to avert such situations in cases when the existence of a local extremum point may be expected, it is possible to adopt a certain modification of the algorithm. Once an extremum has been found, the algorithm chooses a large random displacement which defines a new starting point for another search sequence. Such an operation may be repeated a prescribed number of times before an extremum point is accepted as the solution.

In Fig. 5.14 an example is shown of a successful convergence to the global extremum, in spite of the fact that a local extremum has been located earlier.

The method presented above should be regarded as an example of 'creeping' methods; more such methods can be found in the literature [18, 44, 65].

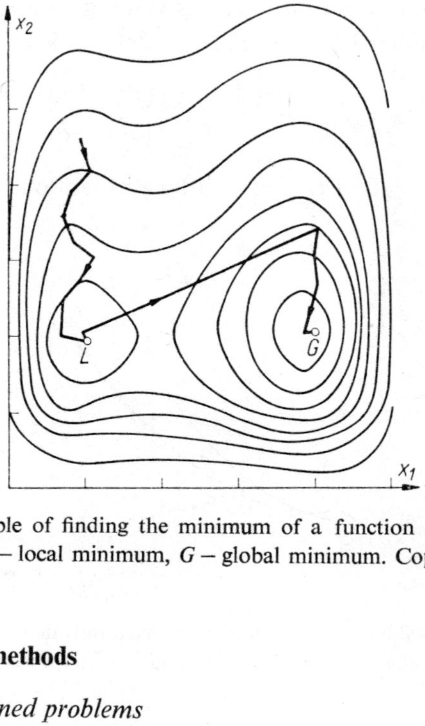

Fig. 5.14 Example of finding the minimum of a function by the random creep method: L – local minimum, G – global minimum. Copyright © 1974 IEEE.

5.3.3 Systematic methods

5.3.3.1 Unconstrained problems

The feature distinguishing the systematic methods is that the series of numerical experiments consists of systematic iterative steps in which the estimates of the extremum are gradually improved. A certain estimate is recognized as the sought extremum if it satisfies a criterion constituting an integral part of the method. The criterion most often relates to certain local properties of the objective function, like the one described in Section 4.5. It is also possible to account for the global properties as recognized in the course of approaching the extremum.

As mentioned in Section 5.3.1, the methods involving searching along the coordinates may fail at the boundary of the feasible region or at a curved ridge of the objective function. The reason is that in these methods the properties of the function are examined only by comparing its values at successively selected points which are distributed along the coordinates. In order to avoid this, the techniques of examination of the function properties are, in certain systematic methods, made independent of the coordinate system by using many search directions, iteratively rotating the coordinate system, etc.

A large group of systematic methods is founded on numerical experiments in which function values only are determined. Such methods are referred to as *derivative-free methods.* Some of these, however, are based on the assumption that the first or even second derivatives of the objective function exist. As a consequence, the class of problems to which such methods can be applied is limited. In other methods the objective function is assumed to be unimodal only, which makes these methods applicable to a very broad class of problems.

The methods based on the assumption that the function is unimodal can be exemplified by the well-known *pattern search* by Hooke and Jeeves [40, 60]. The properties of the objective function are tested by effecting, from the current point, small displacements $\delta_j, j = 1, 2, \ldots, n$ along the coordinate directions. Certain initial magnitudes are assigned to the displacements which are later reduced at specific stages of the search.

Each iterative step consists of the test phase and the pattern move phase. In the following, the points considered in the test phase are denoted as $\boldsymbol{t}$ and those considered in the pattern move as $\boldsymbol{x}$.

At the test phase of the ith step the displacements are effected from the initial point $\boldsymbol{t}_0^{(i)}$ (identical to $\boldsymbol{x}^{(i)}$) according to the following scheme repeated for $j = 1, 2, \ldots, n$

$$\boldsymbol{t}_j^{(i)} = \begin{cases} \boldsymbol{t}_{j-1}^{(i)} + \boldsymbol{\delta}_j & \text{if } f(\boldsymbol{t}_{j-1}^{(i)} + \boldsymbol{\delta}_j) < f(\boldsymbol{t}_{j-1}^{(i)}) \\ \boldsymbol{t}_{j-1}^{(i)} - \boldsymbol{\delta}_j & \text{if } f(\boldsymbol{t}_{j-1}^{(i)} - \boldsymbol{\delta}_j) < f(\boldsymbol{t}_{j-1}^{(i)}) \\ \boldsymbol{t}_{j-1}^{(i)} & \text{otherwise} \end{cases} \tag{5.34}$$

where $\boldsymbol{\delta}_j$ is a vector of length δ_j directed along coordinate x_j. After the test phase has been completed it is checked whether $\boldsymbol{t}_n^{(i)} \neq \boldsymbol{t}_0^{(i)}$, i.e. whether a move has occurred from the initial point. If not, the second phase of the step is bypassed; displacements δ_j are reduced and the next step is initiated by setting $\boldsymbol{x}^{(i+1)} = \boldsymbol{x}^{(i)}$ (and, of course, $\boldsymbol{t}_0^{(i+1)} = \boldsymbol{x}^{(i+1)}$). If the check is positive, then the pattern move is made in the direction established by the preceding successful moves and a new initial point is determined:

$$\boldsymbol{x}^{(i+1)} = \boldsymbol{t}_n^{(i)} + c(\boldsymbol{t}_n^{(i)} - \boldsymbol{x}^{(i)}) \tag{5.35}$$

where c is a constant $\geqslant 1$.

The point defined by (5.35) is accepted if $f(\boldsymbol{x}^{(i+1)}) \leqslant f(\boldsymbol{x}^{(i)})$, otherwise the new initial point is $\boldsymbol{x}^{(i+1)} = \boldsymbol{t}_n^{(i)}$. The search is terminated if it is found in the test phase that after a consecutive reduction the displacements are less than the desired accuracy. An example of application of the pattern search to a function in two variables is shown in Fig. 5.15. The initial estimate of the minimum is $\boldsymbol{x}^{(i)}$; it is additionally assumed in the first step that $\boldsymbol{t}_2^{(0)} = \boldsymbol{x}^{(1)} = \boldsymbol{x}^{(0)}$. In the test phase of the first step a move is made to $\boldsymbol{t}_2^{(1)}$ and in the pattern move phase, to $\boldsymbol{x}^{(2)}$. The second step results in the moves to $\boldsymbol{t}_2^{(2)}$ and $\boldsymbol{x}^{(3)}$. In the third step, the test phase produces a move to $\boldsymbol{t}_2^{(3)}$ and the pattern move phase ends in $\boldsymbol{x}^{(4)}$. In the fourth step, the test phase does not produce any move. In the next step, the initial point remains the same, i.e. $\boldsymbol{x}^{(5)} = \boldsymbol{x}^{(4)}$, while the displacements are reduced; however, the test phase does not produce any move, and so on.

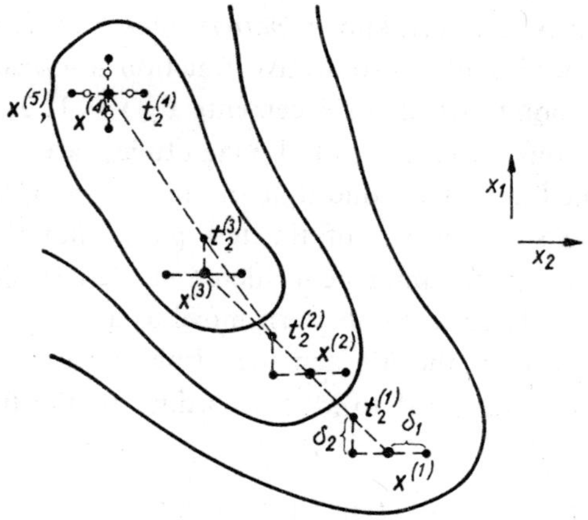

Fig. 5.15 Example of finding the minimum of a function by the pattern search.

The efficiency of the pattern search may change depending on the properties of the function. As the principle of the test phase is arbitrary, the number of function evaluations in one step may vary from n to $2n$.

Another method belonging to the same group of direct search methods is the sequential *simplex search* (not to be confused with Dantzig's simplex algorithm for the problems of linear programming). The simplex in n-dimensional space is a figure of $n+1$ vertices, i.e. on a plane it is a triangle, in three-dimensional space a tetrahedron, in a four-dimensional space a figure of 5 vertices, etc. This technique is based on the examination of the properties of the function by comparing its values at all vertex points of the simplex. The distances between the simplex vertices are assigned certain initial values which may be later reduced as the minimum is approached.

There are several modifications of the sequential simplex algorithm. The most sophisticated one involving changes of the simplex form by contraction or expansion has been devised by Nelder and Mead [40]. An earlier and simpler modification (of Spendley–Hext–Himsworth [45]) is presented below, where each step consists in rejecting one vertex of the simplex and replacing it by its 'mirror image'; in that way the simplex retains its regular shape.

The rules for selecting the point which should be rejected are so adopted that the successively formed figures proceed toward the minimum. The following rules are used:

(1) the vertex at which the function value is highest is rejected;

(2) if applications of rule 1 makes us return to a point which had been rejected in the previous step, the vertex with the second highest value is rejected;

(3) if the best-valued vertex remains unchanged for more than $1.65n$ iterations, the simplex size is reduced; if the distance between the vertices becomes smaller than the required accuracy the search is terminated.

Example 5.6

Use the sequential simplex search to find the minimum of function:

$$f(x_1, x_2) = x_1^2 - 25x_1 + x_2^2 - 2x_2 + 150$$

Let the simplex assume the form of an equilateral triangle situated symmetrically with respect to the straight line inclined at 45° to the axes (Fig. 5.16). It can easily be established that if a is the distance between the simplex vertices then $p = 0.9657a$ and $q = 0.2587a$. If the vertex points are $\boldsymbol{x}^{(A)}$, $\boldsymbol{x}^{(B)}$, $\boldsymbol{x}^{(C)}$ and the point $\boldsymbol{x}^{(A)}$ is to be rejected, then the coordinates of a new vertex point can be calculated as:

$$x_1^{(D)} = x_1^{(B)} + x_1^{(C)} - x_1^{(A)}$$

$$x_2^{(D)} = x_2^{(B)} + x_2^{(C)} - x_2^{(A)}$$

The number of iterations 'approving' the best point (see rule (3)) should be $1.65 \cdot 2 = 3.3$; rounding that value off to the integer gives 4. The process of searching after the minimum for the initial approximation of $x_1^{(0)} = 0$, $x_2^{(0)} = 0$ and distance between the vertices $a = 2$ is shown in Table 5.1 and in Fig. 5.16. The sequence of iterations is interrupted when four successive

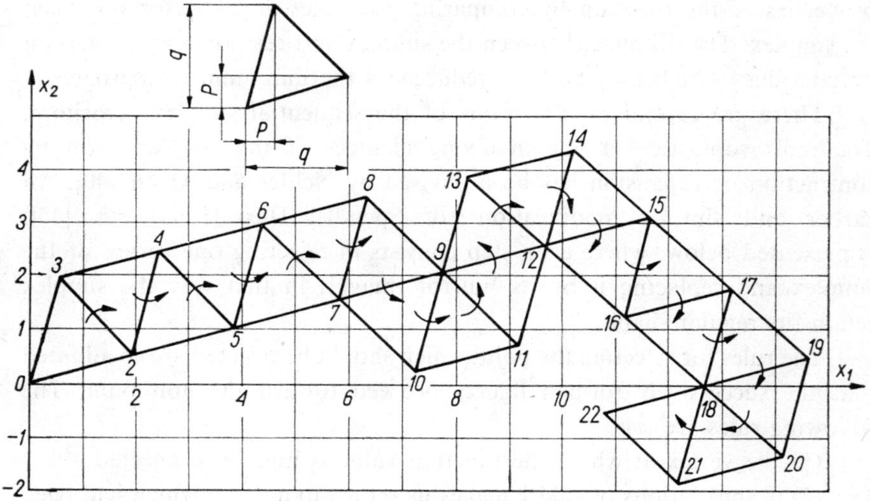

Fig. 5.16 Illustration to Example 5.6 (ref. [45]).

Table 5.1

Search sequence in Example 5.6

Point No.	x_1	x_2	$f(x_1, x_2)$	Points in simplex	Point rejected	Rule for rejection
1	0	0	150.00			
2	1.9314	0.5174	104.68	1 2 3	1	(1)
3	0.5174	1.9314	133.58			
4	2.4488	2.4488	93.78	4 2 3	3	(1)
5	3.8628	1.0348	67.35	4 2 5	2	(1)
6	4.3802	2.9662	64.54	4 6 5	4	(1)
7	5.7942	1.5522	38.02	7 6 5	5	(1)
8	6.3116	3.8436	40.54	7 6 8	6	(1)
9	7.7256	2.0696	16.99	7 9 8	8	(1)
10	7.2082	0.2620	22.30	7 9 10	7	(1)
11	9.2116	0.7794	29.33	11 9 10	10	(2)
12	9.7290	2.5868	2.95	11 9 12	11	(1)
13	8.2430	3.8772	19.15	13 9 12	9	(2)
14	10.2461	4.3944	9.35	13 19 12	13	(1)
15	11.7324	3.1040	−2.34	15 14 12	14	(1)
16	11.2150	1.2964	−5.10	15 16 12	12	(1)
17	13.2184	1.8136	−6.07	15 16 17	15	(1)
18	12.7010	0,0060	−6.22	18 16 17	16	(1)
19	14.7044	0.5232	−2.18	18 19 17	17	(2)
20	14.1870	−1.2840	0.81	18 19 20	19	(2)
21	13.1836	−1.8012	0.69	18 21 20	20	(1)
22	10.6970	−0.5112	−1.71	18 21 22	21	(1)

iterations do not change the 'best' vertex point, which is point No. 18 located at $x_1^{(18)} = 12.7010$ and $x_2^{(18)} = 0.0060$. However, the computations can be continued if the simplex size is reduced. If a is reduced, for example, to 0.2, the next series of iterations leads to the point (12.4319, 1.0105) at which the function value is -7.25.

Another group of methods involving numerical experiments which include the determination of the values of the function and its gradient are referred to as *gradient-based methods*. These methods are applicable to continuously differentiable functions only and, in some cases, require the function to be twice differentiable. As a rule, the convergence to the extremum is faster than in the case of derivative-free techniques.

The archetype of the gradient-based methods is the *method of steepest descent* [49, 60]. The formulae used in this method are given below in vector notation. The principal part of an iterative step which begins at point $\boldsymbol{x}^{(i)}$ is the determination of the function gradient:

$$\nabla f_i = \boldsymbol{k}_1 \frac{\partial f(\boldsymbol{x}^{(i)})}{\partial x_1} + \boldsymbol{k}_2 \frac{\partial f(\boldsymbol{x}^{(i)})}{\partial x_2} + \ldots + \boldsymbol{k}_n \frac{\partial f(\boldsymbol{x}^{(i)})}{\partial x_n} \tag{5.36}$$

where: $\boldsymbol{k}_1, \boldsymbol{k}_2, \ldots, \boldsymbol{k}_n$ are unit vectors along the coordinate axes. The basic idea here is to approach the minimum by moving along the direction of the negative gradient, i.e. the direction vector is $\boldsymbol{p}^{(i)} = -\nabla f_i$. The step length is chosen so as to attain, at least approximately, the lowest value of the function in that direction. In order to reduce the computational effort the function is usually approximated in the direction $\boldsymbol{p}^{(i)}$ by a second or third order polynomial which is denoted below as $y(\boldsymbol{x})$. The necessary calculations consist in solving (with respect to l_i) the equation which presents the necessary condition for the existence of the minimum of $y(\boldsymbol{x})$ in the direction $\boldsymbol{p}^{(i)}$:

$$\boldsymbol{p}^{(i)} \nabla y(\boldsymbol{x}^{(i)} + l_i \boldsymbol{p}^{(i)}) = 0 \tag{5.37}$$

The next approximation of the minimum is determined as:

$$\boldsymbol{x}^{(i+1)} = \boldsymbol{x}^{(i)} + l_i \boldsymbol{p}^{(i)} \tag{5.38}$$

The procedure stops when the step length l_i in successive iterations becomes smaller than the required accuracy.

The classical method of steepest descent is efficient only if the objective function does not exhibit very narrow ravines which may cause characteristic oscillations of the successive steps (Fig. 5.17). The attempts to protect the method from unfavourable function properties resulted in numerous modifications which can be pursued in the literature [6, 48, 49].

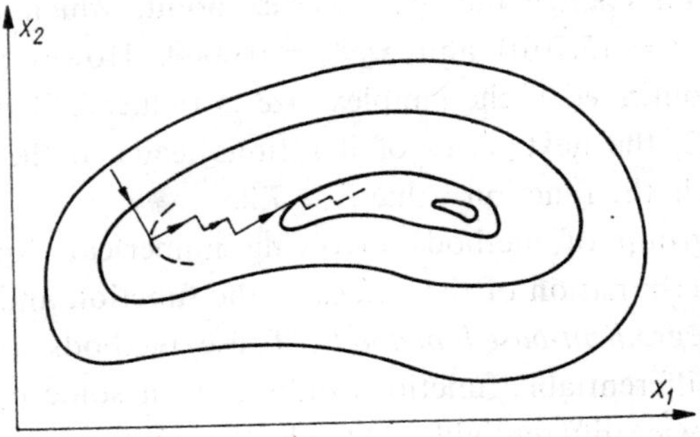

Fig. 5.17 Example of a search sequence by the method of steepest descent in the case of a function of two variables featuring a narrow ravine.

Example 5.7

Use the method of steepest descent to find the minimum of the function:

$$f(x_1, x_2) = x_1^2+x_2^2+1.2x_1x_2$$

The function derivatives assume the form:

$$\frac{\partial f(x_1, x_2)}{\partial x_1} = 2x_1+1.2x_2$$

$$\frac{\partial f(x_1, x_2)}{\partial x_2} = 2x_2+1.2x_1$$

The components of the direction vector may thus be expressed as:

$$p_1 = -2x_1-1.2x_2$$

$$p_2 = -2x_2-1.2x_1$$

Owing to the simple form of the function, it is possible to use in equation (5.37) the original function $f(x_1, x_2)$ instead of the approximating function $y(x_1, x_2)$. Equation (5.37) leads then to the following expression:

$$\frac{\mathrm{d}}{\mathrm{d}l_i} f(x_1^{(i)}+l_i p_1^{(i)}, x_2^{(i)}+l_i p_2^{(i)}) = 0$$

Let the initial variable settings be $x_1^{(0)} = 1$, $x_2^{(0)} = 2$. The sequence of searching after the minimum is presented in Table 5.2 and Fig. 5.18. (It is interrupted after the third step.) The true minimum can of course be found analytically, using the indirect approach: it is the point (0, 0) at which the function value is 0.

The main weakness of the gradient-based methods based on the method of steepest descent is the necessity to rely on gradient properties even in the vicinity of the solution. However, the first derivatives tend to

Table 5.2

Search sequence in Example 5.7

Coordinates of point $x^{(0)}$	1	2
Components of vector $p^{(0)}$	-4.4	-5.2
Components of $x^{(0)}+l_0 p^{(0)}$	$1-4.4\ l_0$	$2-5.2\ l_0$
$f(x^{(0)}+p^{(0)}l_0)$	$82.9\ l_0^2-46.4\ l_0+7.4$	
$(d/dl_0)f(x^{(0)}+l_0\ p^{(0)})$	$165.8\ l_0-46.4$	
l_0	0.280	
Coordinates of point $x^{(1)}$	-0.230	0.543
Components of vector $p^{(1)}$	0.192	-0.808
Components of $x^{(1)}+l_1 p^{(1)}$	$-0.230+0.192\ l_1$	$0.543-0.808\ l_1$
$f(x^{(1)}+l_1\ p^{(1)})$	$0.502\ l_1^2-0.628\ l_1+0.497$	
$(d/dl_1)f(x^{(1)}+l_1\ p^{(1)})$	$1.004\ l_1-0.628$	
l_1	0.627	
Coordinates of point $x^{(2)}$	-0.110	0.038
Components of vector $p^{(2)}$	0.1744	0.0560
Components of $x^{(2)}+l_2 p^{(2)}$	$-0.110+0.1744\ l_2$	$0.038+0.0560\ l_2$
$f(x^{(2)}+l_2\ p^{(2)})$	$0.0452\ l_2^2-0.0336\ l_2+0.0085$	
$(d/dl_2)f(x^{(2)}+l_2\ p^{(2)})$	$0.0904\ l_2-0.0336$	
l_2	0.372	
Coordinates of point $x^{(3)}$	-0.045	0.059
⋮		

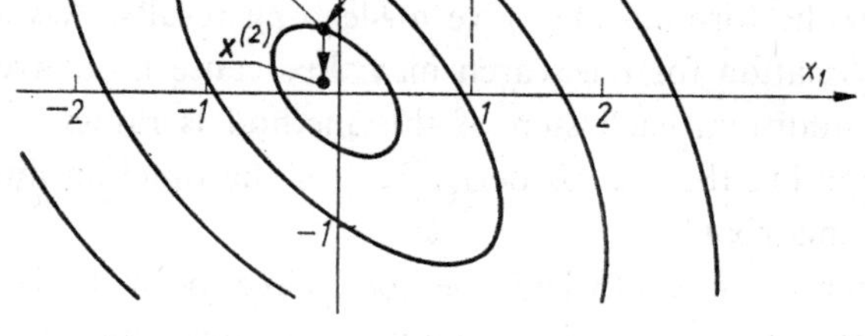

Fig. 5.18 Illustration to Example 5.7.

zero in this region and the behaviour of the function depends on the derivatives of higher order. Another group of so-called *Newtonian methods* [16, 49] has been devised to cope with such difficulties by using additionally second-order derivatives for minimum location. The basic idea employed in such methods originates from the Taylor series expansion of the objective function at the current search point $\boldsymbol{x}^{(i)}$, the derivatives of the third and higher orders being ignored in the series. Using vector notation, the series expansion is:

$$f(\boldsymbol{x}^{(i+1)}) = f(\boldsymbol{x}^{(i)}) + \nabla f_i^T(\boldsymbol{x}^{(i+1)} - \boldsymbol{x}^{(i)}) + \frac{1}{2}(\boldsymbol{x}^{(i+1)} - \boldsymbol{x}^{(i)})^T \mathbf{G}(\boldsymbol{x}^{(i)}) (\boldsymbol{x}^{(i+1)} - \boldsymbol{x}^{(i)}) \tag{5.39}$$

where superscript $\boldsymbol{T}$ denotes the transposed (row) vector, and $\mathbf{G}(\boldsymbol{x}^{(i)})$ is the matrix of second derivatives:

$$\mathbf{G} = \begin{bmatrix} \dfrac{\partial^2 f}{\partial x_1^2} & \dfrac{\partial^2 f}{\partial x_1\,\partial x_2} & \cdots & \dfrac{\partial^2 f}{\partial x_1\,\partial x_n} \\ \dfrac{\partial^2 f}{\partial x_2\,\partial x_1} & \dfrac{\partial^2 f}{\partial x_2^2} & \cdots & \dfrac{\partial^2 f}{\partial x_2\,\partial x_n} \\ \vdots & \vdots & & \vdots \\ \dfrac{\partial^2 f}{\partial x_n\,\partial x_1} & \cdots\cdots & \cdots\cdots & \dfrac{\partial^2 f}{\partial x_n^2} \end{bmatrix} \tag{5.40}$$

It follows from equation (5.39) that the gradient may be expressed as:

$$\nabla f_{i+1} = \nabla f_i + \mathbf{G}(\boldsymbol{x}^{(i)})(\boldsymbol{x}^{(i+1)} - \boldsymbol{x}^{(i)}) \tag{5.41}$$

If the minimum is expected at the point $\boldsymbol{x}^{(i+1)}$, the gradient ∇f_{i+1} should be zero, so that

$$\boldsymbol{x}^{(i+1)} = \boldsymbol{x}^{(i)} - \mathbf{G}^{-1}(\boldsymbol{x}^{(i)}) \nabla f_i \tag{5.42}$$

where $\mathbf{G}^{-1}$ denotes the inverse of the matrix $\mathbf{G}$.

Expression (5.42) is the iterative formula of the *Newton search*. The latter has a very good convergence in the vicinity of the minimum, but outside this region, where expression (5.39) and the neglected terms may be significant, the formula may give misleading results. A very inaccurate initial approximation for the search may thus cause no convergence to be obtained. In addition, each step of this method is rather expensive as it involves computing the second derivatives of the function and finding the inverse of the matrix $\mathbf{G}$.

A number of very efficient methods have been derived from the Newton method by using the idea that near the minimum, the second-order terms of the Taylor series expansion dominate, and, hence, to have a rapidly convergent method for a general function the method must be very fast when applied to a quadratic function. These methods are known as variable metric and conjugate direction methods [16, 26, 40, 90]. Methods of both types minimize a positive definite quadratic function in a maximum of n steps where n is the dimension of the space X. The variable metric achieves this by choosing directions (along which f is to be minimized) which are close to the Newton directions. For a quadratic function, the variable metric direction in the nth step becomes the Newton direction and leads to the minimum in that step. The conjugate directions methods, on the other hand, pick directions $\boldsymbol{p}^{(i)}$ which satisfy the relation:

$$\boldsymbol{p}^{(i)}\mathbf{G}(\boldsymbol{x}^{(i)})\boldsymbol{p}^{(i-1)} = 0$$

For a positive-definite quadratic function, the first n such directions are linearly independent and by a sequential search along these directions the minimum is reached in the nth step.

5.3.3.2 Constrained problems

If the objective function and the constraint system do exhibit certain special features, appropriate systematic methods are available which make it possible to determine the extremum in a finite number of iterative steps. Such possibilities arise when the search algorithm may rely on such features of the problem as the convexity of the region of feasible solutions, definite manner of estimating the function from above or below, etc. Although such special features are seldom found in optimization problems related to process equipment, they are worth mentioning here:

(1) *Linear programming*, where the problems feature a linear objective function and linear constraints. Programs or program packages for solving such problems are available for practically all computers. The possibility of applying the methods of linear programming is thus equivalent to the possibility of performing the computations with the aid of well-proven computational tools. When solving a nonlinear optimization problem, it is always advisable to check whether it could be transformed to a linear form using the techniques presented in Chapter 4 (see also Examples 4.4 and 8.1 as well as the problem presented in Section 11.1).

The methods which are used most often for solving problems of linear programming are [19, 40, 43]:

– the simplex method (of Dantzig) and its modifications,
– the duplex method,
– the Dantzig–Wolfe decomposition method.

When solving linear problems in discrete variables, the methods of Gomory and Land–Doig [33] can be used.

(2) *Quadratic programming*, where the problems feature a quadratic form as the objective function, while the constraints are linear. Computer programs are almost as widely available as in the case of linear programming, so this type of problems is also amenable to numerical solution. The most widely used methods of solving quadratic programming problems are [40, 43]:

– the *Wolfe method*, based on the analogy to linear programming, and leading to the application of the simplex method;

– the *Beale method*, using the Kuhn–Tucker conditions which are exceptionally handy in this case due to the simple form of the problem.

(3) *Separable programming*, where the problems feature a separable objective function and separable constraints. The advantages of the separable form of the objective function have been discussed in Section 4.4. The algorithms for solving such problems rely on the possibility of a piecewise linear approximation of separable functions. For IBM 360/370 computers a system known as MPS (Mathematical Programming System) is available for separable programming problems [50].

(4) *Geometric programming*, where the objective function and the constraints can be expressed as weighted sums of products of variables which are raised to real powers (such expressions are known as posynomials). More precisely, if the problem is given in the standard form:

$$\min_{x} b_0(\boldsymbol{x}) \qquad \boldsymbol{x} = (x_1, x_2, \ldots, x_n), \qquad x_i > 0$$

$$\text{subject to } b_j(\boldsymbol{x}) \leqslant 1 \qquad j = 1, 2, \ldots, m$$

then each of the above functions can be expressed as:

$$b_j(\boldsymbol{x}) = \sum_{k=1}^{K_j} c_{jk} \prod_{l=1}^{n} x_l^{a_{jkl}}, \qquad j = 0, 1, \ldots, m$$

where: K_j is the number of terms in function b_j, c_{jk} are the positively-valued coefficients, and a_{jkl} are arbitrary coefficients. The above problem can be transformed into an equivalent (dual) problem in which we seek the maximum of a function in K variables, where $K = \sum_{j=0}^{m} K_j$. The constraint set in the dual problem includes $n+1$ linear equations and m inequalities.

The existence of the dual problem and its particular features make the problems of geometric programming markedly different from other types of mathematical programming problems [91]. The method of solution depends on the value of $D = K-(n+1)$ known as the *number of degrees of difficulty*. If $D = 0$, the number of variables equals the number of linear equations which can be immediately solved for the sought-after optimal values of the variables. If $D > 0$, the dual problems has to be solved as a problem of finding the constrained extremum – analytically or numerically. The particular features of the posynomials allow the formulation of specialized algorithms which are published in the literature, including programs in FORTRAN [40]. Geometric programming has numerous applications in the field of chemical engineering, predominantly in process optimization [3]. The applications to process equipment design are rather unusual, but a number of optimization problems in general mechanical design have been solved using geometric programming [72].

Systematic methods which are applicable to general problems (i.e. problems incompatible with any of the above special cases) of finding the constrained extremum of a function can be classified into two different groups:

– methods based on the analysis of the feasible region,
– methods using so-called penalty functions.

The first group includes methods which consist of finding search directions allowing the algorithm to remain within the feasible region, or repetitively performing the search on a sequence of well-defined sets containing the feasible region. Methods belonging to this group include the method of feasible directions, the gradient projection method and the cutting plane method. More information on those methods can be found in the literature [26, 40, 48, 49, 90].

The second group of methods has gained popularity owing to the fact that these allow for performing computations using algorithms suited to unconstrained problems. This is possible because the basic idea of such methods is to transform the original problem into a series of unconstrained problems with the property that the series of solutions converges to the sought extremum [15, 16, 26, 40, 48, 49, 60].

The method employing the so-called outside (exterior) penalty function can be applied only if the objective function is mathematically defined outside the feasible region. An outside penalty function is one which is added to the objective function and which adversely affects the value of the combined function when an associated constraint is violated. A penalty function $K(\boldsymbol{x})$ must satisfy the following requirements:

– it must be continuously differentiable,
– $K(\boldsymbol{x}) = 0$ for all $\boldsymbol{x}$ satisfying the constraints,
– $K(\boldsymbol{x}) > 0$ for all $\boldsymbol{x}$ violating the constraints.

The basic idea of the methods discussed here is to find, for an ascending series of positively-valued numbers $r_1, r_2, \ldots, r_i, \ldots$, a series of solutions to the following unconstrained problems:

$$\min_{x} P_i(x) = f(x) + r_i K(x) \tag{5.43}$$

The usefulness of the above transformations lies in the fact that when r_i increases without limit, the series of solutions of problems (5.43) converge to the solution of the problem of minimizing $f(x)$ with constraints. This is additionally explained in two figures. As the term $r_i K(x)$ increases in value with increasing i, Fig. 5.19, then the minimum of function $f(x) + r_i K(x)$ moves towards constrained minimum of $f(x)$, Fig. 5.20.

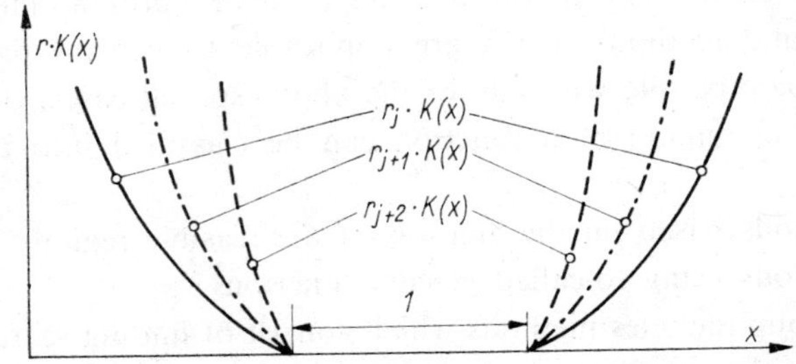

Fig. 5.19 Changes in the term $rK(x)$ in a series of problems including outside penalty function: 1 – feasible region.

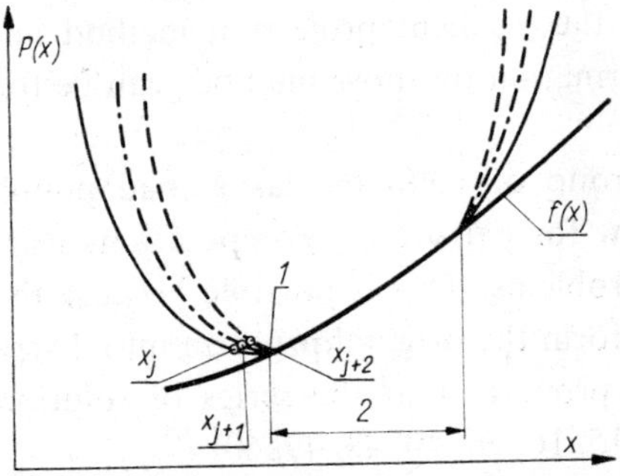

Fig. 5.20 Solutions of a series of problems including outside penalty function: 1 – constrained minimum, 2 – feasible region.

Various methods using outside penalty functions which can be found in the literature differ with respect to the manner in which the series of coefficients r_i are defined and also with respect to the definitions of the penalty functions. If the constraint system consists of inequalities of the form:

$$h_j(x) \leqslant 0, \qquad j = 1, 2, \ldots q \tag{5.44}$$

the penalty function can be defined as, for example:

$$K(x) = \sum_{j=1}^{q} [h_j(x)+|h_j(x)|]^2 \tag{5.45}$$

$$K(x) = \sum_{j=1}^{q} \max[0, h_j(x)] \tag{5.46}$$

Note that if the problem includes equality constraints of the form $g_j(x) = 0$, it is possible to transform each of them to a pair of inequalities:

$$g_j(x) \leqslant 0 \qquad -g_j(x) \leqslant 0 \tag{5.47}$$

and then use one of the formulae given above. The effect of the penalty function is additionally illustrated in Fig. 5.21.

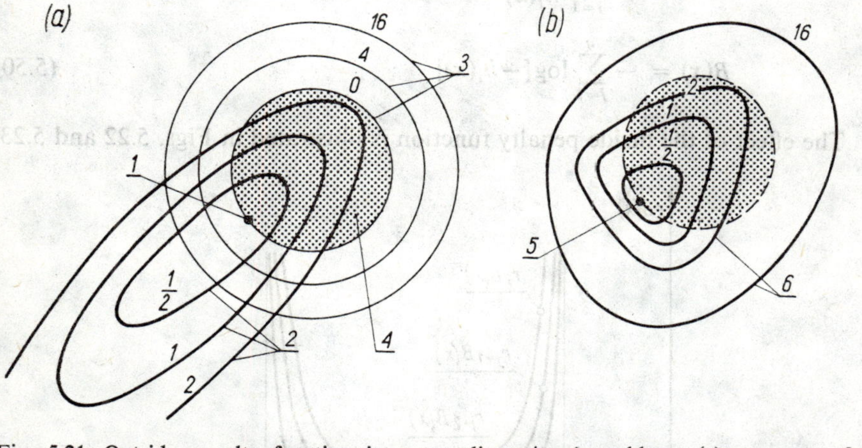

Fig. 5.21 Outside penalty function in a two-dimensional problem: (a) contours of objective function f and outside penalty function K; (b) contours of function P as defined by (5.43). 1 – minimum of $f(x)$ in feasible region, 2 – contours of $f(x)$, 3 – contours of $K(x)$, 4 – $K(x) = 0$ in feasible region, 5 – minimum of $P(x)$, 6 – contours of $P(x)$.

The methods employing the so-called inside (interior) penalty function may be applied as an alternative to outside penalty function methods. However, the class of problems to which inside penalty functions can be applied is larger and includes also problems featuring an objective function which is undefined mathematically outside the feasible region. An inside penalty function $B(x)$ must satisfy the following requirements:

– it must be continuously differentiable,
– it increases without limit when x approaches the border of the feasible region.

The inside penalty function is used to construct for a descending series of positively-valued numbers $r_1, r_2, \ldots, r_i, \ldots$, a series of unconstrained problems:

$$\min_{x} P_i(x) = f(x) + r_i B(x) \tag{5.48}$$

The idea underlying this transformation is that when r_i tends to zero, the series of solutions of problems (5.48) converges to the constrained minimum of $f(x)$.

Various algorithms for constructing the series of coefficients r_i and various definitions of inside penalty functions can be found in the literature. Two possible definitions are as follows:

$$B(x) = -\sum_{j=1}^{q} \frac{1}{h_j(x)} \tag{5.49}$$

$$B(x) = -\sum_{j=1}^{q} \log[-h_j(x)] \tag{5.50}$$

The effect of the inside penalty function is illustrated in Figs. 5.22 and 5.23.

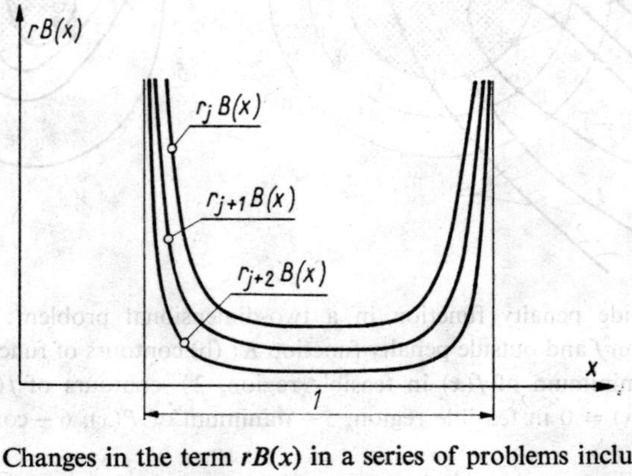

Fig. 5.22 Changes in the term $rB(x)$ in a series of problems including inside penalty function: 1 – feasible region.

As already mentioned in Section 4.2, penalty function methods may be used to transform an optimization problem in discrete variables into a series of problems in continuous variables. Consider a problem including constraints of the form (5.44), while the variables $x_1, x_2, \ldots, x_d$ are discrete and the remaining variables are continuous (a problem can be also imagined with discrete variables only, i.e. $d = n$). The idea behind transformation into a series of unconstrained problems is to define two ascending series of positively-valued numbers $r_1, r_2, \ldots, r_i, \ldots$ and $s_1, s_2, \ldots, s_i, \ldots$ and to minimize the expression

$$P_i(x) \doteq f(x) + r_i K(x) + s_i Q(x_1, x_2, \ldots, x_d) \tag{5.51}$$

where: $K(x)$ is the outside or inside penalty function defined by one of the formulae (5.45), (5.46), (5.49) and (5.50); $Q(x_1, x_2, \ldots, x_d)$ is a special function which penalizes any departure from the permissible values of discrete variables. The special function must satisfy requirements which are similar to those governing $K(x)$:

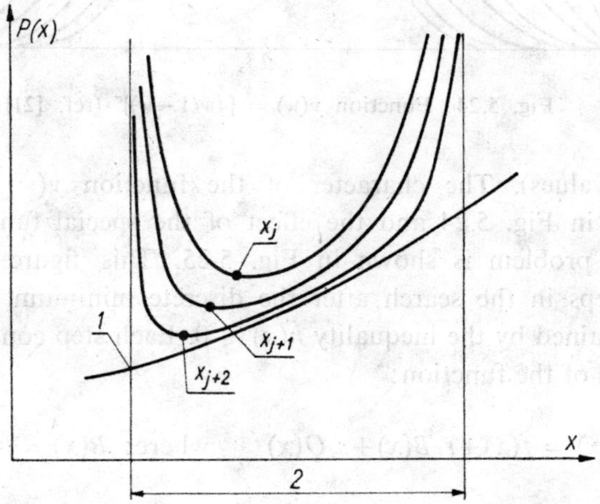

Fig. 5.23 Solutions of a series of problems including inside penalty function: 1 – minimum, 2 – feasible region.

– it must be continuously differentiable,

– $Q(x_1, x_2, \ldots, x_d) = 0$ for values $x_1, x_2, \ldots, x_d$ belonging to the corresponding sets of permissible values,

– $Q(x_1, x_2, \ldots, x_d) > 0$ for 'forbidden' values of $x_1, x_2, \ldots, x_d$.

The function may assume the form:

$$Q(x_1, x_2, \ldots, x_d) = \sum_{j=1}^{d} [4w_j(1 - w_j)]^v \tag{5.52}$$

where: $v \geqslant 1$, and w_j is defined by the expression:

$$w_j = \frac{x_j - z_j^{(l)}}{z_j^{(r)} - z_j^{(l)}} \tag{5.53}$$

while $z_j^{(l)}$ and $z_j^{(r)}$ are the permissible values of the variable x_j which are closest to its current value (superscripts l, r denote the 'left' and 'right'

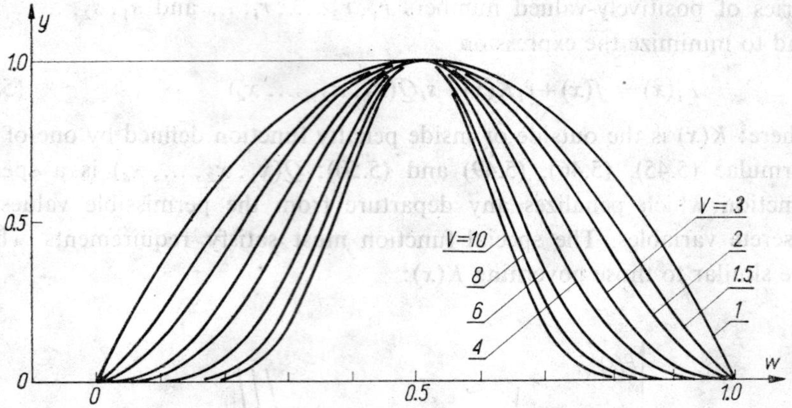

Fig. 5.24 Function $y(w) = [4w(1-w)]^v$ (ref. [21]).

permissible values). The character of the function $y(w) = [4w(1-w)]^v$ is illustrated in Fig. 5.24 and the effect of the special function Q in an optimization problem is shown in Fig. 5.25. This figure depicts three successive steps in the search after the discrete minimum of $f(x)$ in the region constrained by the inequality $h(x) \leqslant 0$. Each step consists in finding the minimum of the function:

$$P_i(x) = f(x) + r_i B(x) + s_i Q(x) \qquad \text{where: } B(x) = -\frac{1}{h(x)}$$

The function $Q(x)$ is defined according to (5.52) and (5.53), $d = 1$, and the permissible values of x are a, b, c, d. Owing to the fact that $r_1 > r_2 > r_3$ and $s_1 < s_2 < s_3$, the series of x values determined by successive minimizations of $P_i(x)$, i.e. $x^{(1)}, x^{(2)}, x^{(3)}, \ldots$ converge to the true discrete minimum $x = b$. For information on how the series of coefficients r_i and s_i should be constructed, and for comments on the practical applications of the method described, see [21].

5.3.4 Combined methods

This group of methods combines the features of systematic and random search methods. The random element is introduced in an attempt to increase, in the case when local extremum points may exist, the chance that the global extremum will be found. This makes the methods useful in situations where the properties of the objective function have not been fully recognized.

The series of numerical experiments performed according to combined methods consist of iterative steps as with systematic methods. The series is repeated several times, each time yielding an extremum which is identified

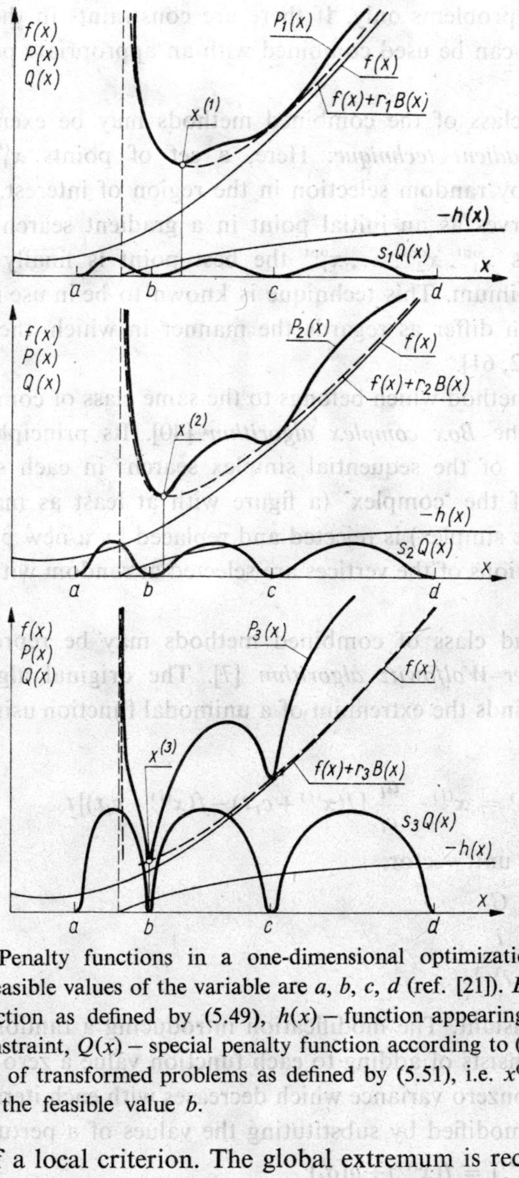

Fig. 5.25 Penalty functions in a one-dimensional optimization problem where the feasible values of the variable are a, b, c, d (ref. [21]). $B(x)$ – inside penalty function as defined by (5.49), $h(x)$ – function appearing in the inequality constraint, $Q(x)$ – special penalty function according to (5.52). Series of solutions of transformed problems as defined by (5.51), i.e. $x^{(1)}$, $x^{(2)}$, $x^{(3)}$ approaches the feasible value b.

with the aid of a local criterion. The global extremum is recognized as the best among the points discovered. The random element may be introduced into the search in two different ways:

(1) by selecting the initial points at random for each series,

(2) by employing a random search in each series.

A short review of combined methods given below includes one technique for constrained minimization (Box algorithm) and two methods suited to unconstrained problems only. If there are constraints in the problem, the latter methods can be used combined with an appropriate penalty function technique.

The first class of the combined methods may be exemplified by the *randomized gradient technique*. Here, a set of points $\boldsymbol{x}_1^{(0)}, \boldsymbol{x}_2^{(0)}, \ldots, \boldsymbol{x}_r^{(0)}$ is established by random selection in the region of interest. Each of these points then serves as an initial point in a gradient search. In the set of terminal points $\boldsymbol{x}_1^{\text{opt}}, \boldsymbol{x}_2^{\text{opt}}, \ldots, \boldsymbol{x}_r^{\text{opt}}$ the best point is finally recognized as the global minimum. This technique is known to be in use in several modifications which differ as regards the manner in which the initial points are selected [22, 61].

Another method which belongs to the same class of combined methods is known as the *Box complex algorithm* [40]. Its principle is somewhat similar to that of the sequential simplex search: in each step, the worst vertex point of the 'complex' (a figure with at least as many vertices as there are in the simplex) is rejected and replaced by a new point. However, the initial positions of the vertices are selected at random within the feasible region.

The second class of combined methods may be represented by the modified *Kiefer–Wolfowitz algorithm* [7]. The original algorithm, which is systematic, finds the extremum of a unimodal function using the iteration formula

$$\boldsymbol{x}^{(i+1)} = \boldsymbol{x}^{(i)} - \frac{a_i}{c_i}\left[f(\boldsymbol{x}^{(i)} + c_i \boldsymbol{t}) - f(\boldsymbol{x}^{(i)} - c_i \boldsymbol{t})\right]\boldsymbol{t} \tag{5.54}$$

where: $\boldsymbol{t}$ is the unit vector,

$$a_i = \frac{G}{i} \tag{5.55}$$

$$c_i = i^{1/3} \tag{5.56}$$

and G is a constant. The modification introducing a random element into the method consists of adding to each function value a zero mean artificial noise with a nonzero variance which decreases with each iteration. Formula (5.54) is thus modified by substituting the values of a perturbed function:

$$f^*(\boldsymbol{x}^{(i)}) = f(\boldsymbol{x}^{(i)}) + p(\sigma_i) \tag{5.57}$$

while the artificial perturbation $p(\sigma_i)$ must satisfy two conditions:

$$\mathscr{E}[p(\sigma_i)] = 0 \tag{5.58}$$

$$\lim_{i \to \infty} \mathscr{E}[p^2(\sigma_i)] = 0 \tag{5.59}$$

where $\mathscr{E}$ denotes the expected value operator.

A Gaussian perturbation may be used here as it satisfies the above conditions, i.e.

$$p(\sigma_i) = \frac{1}{\sqrt{2\pi}\,\sigma_i} \exp\left[-\frac{1}{2}\left(\frac{y}{\sigma_i}\right)^2\right] \tag{5.60}$$

where: $\sigma_i = \dfrac{K}{i}$, K is a constant, and y a random variable. The values of G and K are selected depending on the size of the region to be searched.

The random perturbations used in the modified Kiefer–Wolfowitz algorithm make the trajectories of the search procedure look different even if the searches start at the same initial point. This makes the probability of reaching the global extremum increase with the number of searches, even if the function has local extremum points. Figure 5.26 illustrates two typical search trajectories starting from the same point in the case of the function:

$$f(x_1, x_2) = \left[1 - 8x_1 + \frac{7}{3}(x_1^2 - x_1^3) + \frac{1}{4}x_1^4\right] x_2^2 e^{-x_2}$$

which has two minimum points: a local minimum at (1, 2) and a global one at (4, 2).

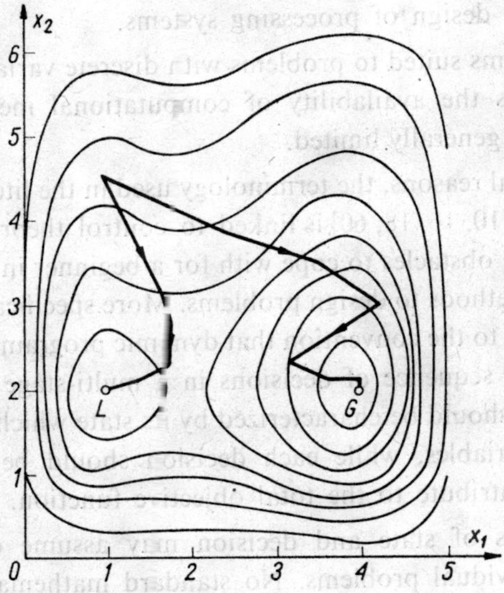

Fig. 5.26 Example of finding the minimum of a function by the modified Kiefer–Wolfowitz algorithm (ref. [7]): L – local minimum, G – global minimum. Copyright © 1974 IEEE.

5.3.5 Method of dynamic programming

Dynamic programming is a mathematical optimization approach used for making a series of interrelated decisions. Its applications are mainly concerned with the economic problems of resource allocation, optimal decisions in process control and optimal design of complex technological systems. The applications to optimization of equipment design include multistage gear trains [22, 72].

A specific feature of the dynamic programming approach is that large optimization problems including many decisions or requiring the checking of a large number of discrete variations are solved as a series of single-stage decision processes. Such an approach to problem formulation should not be confused with the decomposition techniques discussed in Section 4.6.

A major distinction among dynamic programming problems is the nature of the decision variable, and thus two groups of computational methods can be distinguished:

– algorithms suited to problems with continuous variables,

– algorithms suited to problems with discrete variables.

In the technological problems considered in this book the first group of algorithms is seldom used. However, such algorithms are very efficient when applied to sequential problems with a large number of variables appearing in the design of processing systems.

The algorithms suited to problems with discrete variables deserve more attention here as the availability of computational methods for solving such problems is generally limited.

For historical reasons, the terminology used in the literature on dynamic programming [10, 16, 18, 60] is linked to control theory and economics. This is one of the obstacles to cope with for a beginner in the field of application of these methods to design problems. More specifically, it is necessary to fit the problem to the convention that dynamic programming is concerned with the optimal sequence of decisions in a multi-stage decision process. Next, each stage should be characterized by its state which can be expressed through state variables, while each decision should be assigned certain results which contribute to the total objective function.

The concepts of state and decision may assume different physical meanings in individual problems. No standard mathematical formulation exists for dynamic programming problems. What these problems have in common is the recursive character of the formulae expressing the results of decisions in successive stages. This formulation is used to determine the optimal decision for all possible states of the system for the first stage of the problem.

Typically, the first stage in a dynamic programming formulation refers to the last decision which must be made in a series of sequential decisions. The results of this stage are then included in the next stage of the dynamic programming formulation. This recursive procedure is applied at each stage until the last stage is reached, at which point it becomes possible to determine the optimal value for the initial decision of the problem. Finally, the optimal decisions at each stage can be determined by retracing the steps through the set of optimal decisions that were found for each state of the system at previous stages.

There are many computational methods suited to particular features of recursive relationships defining particular problems. All the methods are based on *Bellman's principle of optimality*: in a n-stage optimal decision process leading to the final stage characterized by state S_n, its $(n-p)$-stage part leading from the pth intermediate stage (state S_p) to the final stage (state S_n) must also be optimal in itself. Bellman's principle expressed in common speech says that whatever the initial state and initial decisions are, the decisions applied in the remaining stages must be optimal with respect to the state resulting from the initial decision, if the overall decision process is to be potentially optimal.

As already mentioned, a prerequisite for the application of the dynamic programming approach is the formulation of the objective function in a manner which reflects the idea that the entire problem is treated as a series of single-stage decision processes. The idea consists of relying on a recursive formulation where for any part of the decision process, say from the beginning to the ith stage, the function depends upon: the function value corresponding to the process from the beginning to the $(i-1)$th stage, state of the ith stage and decision taken at the ith stage.

The general dynamic programming recursive formulation may be expressed as follows:

$$f_n(n, S_n, x_n) = F[g(n, S_n, x_n), f^*_{n-1}(S_{n-1})] \tag{5.61}$$

where: f_n, F, g, f^*_{n-1} are the known functions, n is the number of variables and, at the same time, the number of decision stages, S_n is the state of the system at stage n, $g(n, S_n, x_n)$ is the immediate result associated with taking decision x_n when the state of the system is $S_n, f^*_{n-1}(S_{n-1})$ is the result associated with the optimal sequence of decisions up to stage $(n-1)$ when the state is S_{n-1}. From the recursive formulation of the objective function it follows that it is possible to express analogously the evaluation of the decision sequences extending to stages $(n-1)$, $(n-2)$, etc., until the sequence containing the first element only is reached.

The recursive formulation of the objective function can be illustrated by the example of a linear problem with a single state variable (this is a purely speculative example as such problems are usually solved by special methods which are more efficient than dynamic programming, see Section 5.3.3.2). For a linear objective function and a linear constraint

$$\min_x \sum_{i=1}^{n} c_i x_i$$
$$\text{subject to } \sum_{i=1}^{n} a_i x_i - s \leqslant 0 \tag{5.62}$$

the relationship (5.61) assumes the form

$$f_n(n, s, x_n) = c_n x_n + f^*_{n-1}(s - a_n x_n) \tag{5.63}$$

the recursive formulation applied to the next element of the decision sequence yields:

$$f_{n-1}(n-1, s_{n-1}, x_{n-1}) = c_{n-1} x_{n-1} + f^*_{n-2}(s_{n-1} - a_{n-1} x_{n-1})$$

where: $s_{n-1} = s - a_n x_n$, etc.

The recursive formulation of the objective function as expressed by (5.61) comes into play most naturally if the object considered may be interpreted as a chain structure of elements (Fig. 5.27). In special cases, the sequential form may arise for other reasons.

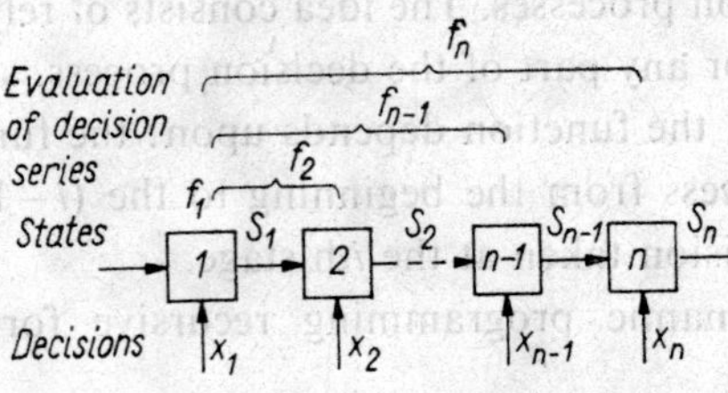

Fig. 5.27 Recursive description of a chain-structured object.

The computational methods for discrete dynamic programming problems involve a two-stage procedure:

(1) For $i = 1, 2, \ldots, n$ the optimal x_i values are found for each state of the next stage using the optimal results from the previous stage.

(2) For $i = n-1, n-2, \ldots, 1$ the sequence of optimal decisions is traced by comparing the values of the objective function and identifying the states which result from previous decisions.

This is indeed a clever scheme for making a search in the set of feasible solutions. Owing to the principle of optimality it is possible to immediately neglect such decisions which certainly do not belong to the sequence of optimal decisions. This makes it possible to avoid keeping in mind many different combinations of decisions and states, together with the corresponding values of the objective function. Consequently, the examination of function values in the second part of the procedure also becomes easier. The logic diagram illustrating the principle of discrete dynamic programming is shown in Fig. 5.28. This principle is used, for example, in Rider's algorithm for the solution of problems with a single state variable; this algorithm has been published as a FORTRAN computer program [40].

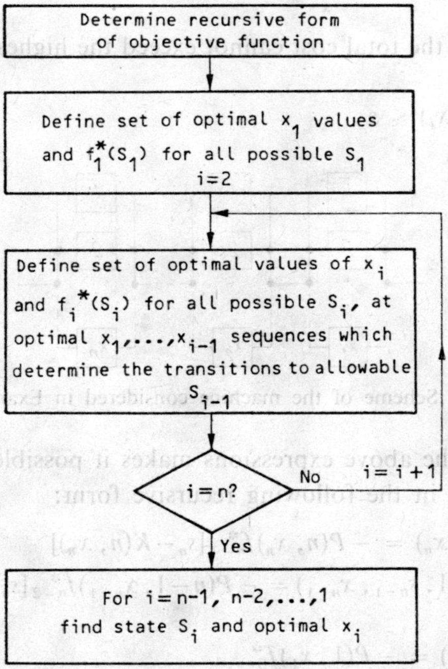

Fig. 5.28 Simplified logic diagram of the algorithm for solving discrete dynamic programming problems.

Example 5.8

Formulate as a dynamic programming problem the search for the structure of a machine giving the highest over-all reliability. The machine consists of n types of serially linked subassemblies, each of which must function if the machine is to function (Fig. 5.29). The probability that the ith subassembly will function is $P(i, 1)$; the probability increases to $P(i, x_i)$ if there are

x_i such subassemblies installed in parallel. The cost of a single ith subassembly is $K(i, 1)$; the cost of x_i parallel subassemblies is $K(i, x_i)$. The total cost of the machine may not exceed the value s.

The above problem can be treated as a n-stage decision process (one stage = one kind of subassembly). The state in each stage is defined by a single variable which may be attributed the meaning of the unallocated part of the allowable cost at a particular stage. The decision variable is the number of parallel units for each subassembly. The highest reliability of the machine is attained if the negative product of probabilities that the stages will function attains its minimum:

$$-\prod_{i=1}^{n} P(i, x_i) \to \min_{x}$$

At the same time, the total cost cannot exceed the highest allowable value:

$$\sum_{i=1}^{n} K(i, x_i) \leqslant s$$

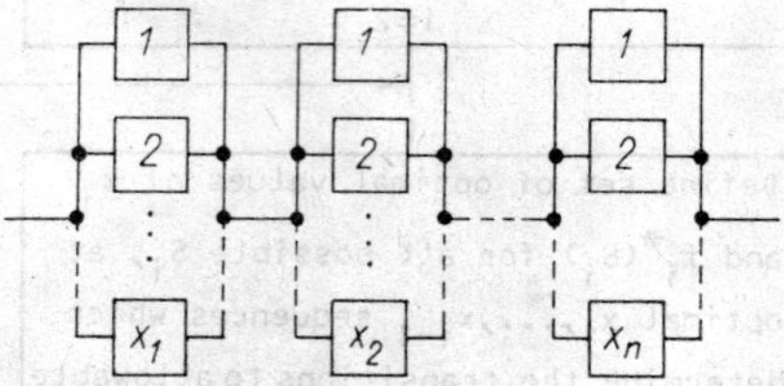

Fig. 5.29 Scheme of the machine considered in Example 5.8.

The structure of the above expressions makes it possible to formulate the objective function in the following recursive form:

$$f_n(n, s_n, x_n) = -P(n, x_n) f_{n-1}^{*}[s_n - K(n, x_n)]$$

$$f_{n-1}(n-1, s_{n-1}, x_{n-1}) = -P(n-1, x_{n-1}) f_{n-2}^{*}[s_{n-1} - K(n-1, x_{n-1})]$$

$$\vdots$$

$$f_1(s_1, x_1) = -P(1, x_1) f_0^{*}$$

where: $s_n = s$, $s_i = s_{i+1} - K(i+1, x_{i+1})$ for $i = 1, 2, \ldots, n-1$, $f_0^{*} = 1$.

The above formulae make it possible to solve the problem using the algorithm according to the logic diagram shown in Fig. 5.28.

5.4 SELECTION OF THE METHOD

Selection of one search technique from many for a given optimization problem has two fundamental aspects:

— it must be possible to obtain a correct solution,

– the computation effort or the cost of computations must be reduced to minimum.

The first aspect requires a determination of the mathematical properties of the problem as completely as possible and verification of the compatibility of the method with those properties. This would be a relatively simple task, were it not frequently necessary to approach the solving of a problem without knowing much about its properties, especially when their determination turns out to be a difficult problem in itself.

In general, it is fairly easy to establish whether the problem can be solved by indirect methods, as that depends only on the existence and the continuity of the function and possibly its derivatives. On the other hand, it may happen that the selection of one direct method from many cannot be thoroughly analysed and appropriately decided. In particular, it is sometimes virtually impossible to recognize all important properties of the constraint system or to establish the existence of local extremum points. If the mathematical nature of the problem is not fully understood, it becomes necessary to select a search technique which is suited to the worst possible case. Even if our knowledge of the problem were complete the selection of a specific method for its solution might not necessarily be completely straightforward. From experience we know that, given a specific search technique, it is generally possible to find an objective function and an initial search point for which the given search technique performs at least as well as any other. It follows that comparisons of direct methods – usually made on the basis of their performance in one or in several test problems – are not very reliable (for reviews of test problems see [60, 89]).

In spite of all the uncertainties, the selection of a direct method for finding the extremum should depend on:

(1) the requirements on the computational equipment following from the features of the algorithm;

(2) the experience in applying the method to problems of the same type;

(3) the required accuracy of determination of the results;

(4) the estimated cost of the computations;

(5) the relationship between the time spent on evaluating the objective function and its derivatives, if used, and the time spent on other aspects of the search.

Considering the first factor, one can possibly bring the selection of a method into line with the potential of the available computational equipment. There are certain methods which involve complicated algorithms of determination of the function derivatives, inversion of matrices, etc. Such methods are practically realizable only on a large computer. On the other hand, certain simple algorithms are well suited – particularly in problems with a few variables only – to small desk-top computers or programmable calculators. The practical importance of the factor (2) and (3) is self evident and, in particular, the suitability of the method may depend on the desired accuracy. The last two factors may affect the selection if it follows from the preceding considerations that several different methods can be applied to a given problem.

The factor (5) touches an important question of the numerical properties of the computational methods in which function derivatives are used. Namely, some algorithms require analytical expressions to be available for the computation of the derivatives; in other algorithms, the derivative values are determined by numerical differentiation. In the first case, particularly in problems involving many variables and a complicated objective function, a considerable amount of work is needed to find analytical expressions defining function derivatives. In the second case, the initial treatment is much easier; however, the numerical differentiation of certain functions may require a computational effort substantially greater than required to compute the derivatives from analytical formulae. On the other hand, certain practical problems involve objective functions for which no analytical formulae are available; then, the methods of the first type cannot be used at all while the other type may give excellent results. This is a fairly frequent occurence with mathematical models used in design of process equipment. These could involve complicated algorithms in which, for example, numerical solution of transcendental equations may occur. Several examples of such models can be found in Chapters 8 and 10.

If consideration is being given to methods available in the form of computer programs, the approach to the selection should be modified with the emphasis on factor (2). It should be observed, however, that the application of a particular computer program constitutes a problem very different from the application of a method in general. The quality of computer programs, even those incorporated in software libraries offered by well-known computer companies, may differ considerably. As a result, different computer programs using the same search method may exhibit different efficiencies even if applied to the same test problem [1, 13].

Example 5.9

Two computer programs are available for finding the unconstrained minimum of a function: one using the sequential simplex search and another using the univariate search based on the quadratic interpolation method in one dimension. Select the program which is better suited to finding the values of variables x_1 to x_4 which minimize the function

$$f(x) = \sum_{j=1}^{m} (x_1 + x_2 a_j + x_2 x_3 a_j^2 + x_1 x_4 a_j^3 - b_j)^2$$

where: m is a given number of the order 100 and a_j, b_j are known numbers; $a_j < 1$; $x_i > 0$.

As the function is continuously differentiable, the problem could actually be solved by an indirect method; however, the initial treatment would be time-consuming and a considerable computational effort would be required because the form of the derivatives is rather inconvenient. As to the direct methods available, both can be easily applied although with different efficiency. The sequential simplex method would require more iterations than the interpolation-based univariate search. The latter method involves quadratic interpolation which fits the actual function accurately and thus the convergence becomes faster as the successive approximations approach the solution. As a consequence, the program using the univariate search is preferable in this case. Obviously, the above conclusion holds only if the two programs are of similar quality.

CHAPTER 6

Interpretation of optimization results

6.1 VERIFICATION OF PROBLEM FORMULATION

The formulation of the problem may require resolving various questions which arise from an incomplete theoretical basis for the mathematical modelling, the lack of accurate numerical data, or simply from the lack of solutions to similar problems. The consequences of the assumptions made while formulating the problems are, in certain cases, not apparent until the first results are obtained. If these results indicate that some mistakes have been made, then the initial assumptions must be corrected. Therefore, the work on design optimization cannot be brought to an end, as soon as a solution to the mathematical problem has been obtained. Before the results are accepted as a basis for engineering decisions, their potential consequences must be thoroughly analyzed. Experience shows that such an analysis may sometimes lead to the identification of corrections to be made to the mathematical model, evaluation criterion, or even the method of solution. Actually, the point is to verify the correctness of interpretation of the results, rather than simple mistakes. While problem formulation involves transforming engineering facts into abstract mathematical forms, the analysis of the results is actually a reverse operation, as numerical data and understanding of mathematical relationships are transformed into conclusions of a technological or economic nature. The verification of these conclusions, with general knowledge of the problem and engineering intuition taken into account, seems to be one of the most efficient measures which could be taken to validate the problem formulation. It should be realized, however, that such a procedure also has its drawbacks. If necessary corrections are identified only when controlling the results of the problem solution, then these must be introduced and the computations repeated, which may substantially increase the solution cost.

6.2 SENSITIVITY ANALYSIS

Apart from the verification of problem formulation, the analysis of the optimization results makes it possible to investigate the following questions:

(1) Changes in the solution quality resulting from parameter changes (e.g. at variable process parameters). It should be remembered that the classification of practical problems into deterministic and probabilistic ones is rather arbitrary. As a rule, parameter values are determined with some uncertainty and it may be desirable to evaluate the associated technological or economic risk.

(2) The effect of changes in the final form of the object of the design. This problem is related not only to unavoidable manufacturing errors but also to design changes which may be desirable for a variety of reasons, like ease of part machining or assembling, planned process alterations, etc.

(3) Allowable manufacturing tolerances for the most important equipment parts. The problem may be posed as finding the limits of tolerances with respect to the feasibility of the solution, or as finding the tolerances which do not substantially affect product quality.

The problems mentioned above belong to the scope of so-called *sensitivity analysis.* A system is regarded as sensitive to certain variables when changes in them can substantially affect the system characteristics. Sensitivity should be understood in the global sense, in that all the possible deviations of the quantity (or quantities) considered must be taken into account. Local sensitivity is a slightly different concept and is concerned with small deviations of the quantity (or quantities) considered in the vicinity of a certain point.

In design problems, local sensitivity in the neighbourhood of the optimum is usually investigated. The importance of the methods of sensitivity analysis lies in the fact that they enable one to evaluate the consequences of simultaneous deviations of various quantities. If, however, the deviations of important quantities are being considered separately, then it is difficult to evaluate correctly the effect of some particular combination of deviations.

Sensitivity analysis can be performed by different methods related to the mathematical properties of the optimization problem considered and of the solution method applied. In the case of problems solved by exhaustive search or random search, it may be sufficient to retain several feasible points where the objective function value is close to optimal, and to analyze their locations in relation to the optimum. If the objective function and constraint functions are given analytically and the functions are at least twice differentiable, then the sensitivity analysis can be a rigorous procedure. Let us consider such a procedure in this case.

Mathematically, the sensitivity analysis is based on the Taylor series expansion of the objective function (or an augmented objective function) in the neighbourhood of the optimum:

$$f(\boldsymbol{x}) = f(\boldsymbol{x}^{\text{opt}})+(\boldsymbol{x}-\boldsymbol{x}^{\text{opt}})^T\nabla f(\boldsymbol{x}^{\text{opt}})+\frac{1}{2}(\boldsymbol{x}-\boldsymbol{x}^{\text{opt}})^T\mathbf{G}(\boldsymbol{x}-\boldsymbol{x}^{\text{opt}})+ \ldots \tag{6.1}$$

where $(\boldsymbol{x}-\boldsymbol{x}^{\text{opt}})$ is a column vector and $(\boldsymbol{x}-\boldsymbol{x}^{\text{opt}})^T$ – a row vector of variable deviations; $\nabla f(\boldsymbol{x}^{\text{opt}})$ is the gradient, i.e. column vector of first derivatives, **G** is the matrix of second partial derivatives, also called Hessian matrix of function f, evaluated at $\boldsymbol{x} = \boldsymbol{x}^{\text{opt}}$.

Our considerations will be confined to the case when $\boldsymbol{x}^{\text{opt}}$ is a stationary point, i.e. $\nabla f(\boldsymbol{x}^{\text{opt}})$ is a zero-valued vector. For the sake of brevity, let us introduce the following symbols:

$$\boldsymbol{w} = \left(\frac{x_1-x_1^{\text{opt}}}{x_1^{\text{opt}}}, \frac{x_2-x_2^{\text{opt}}}{x_2^{\text{opt}}}, \ldots, \frac{x_n-x_n^{\text{opt}}}{x_n^{\text{opt}}}\right)^T \tag{6.2}$$

$$\mathbf{L} = \begin{bmatrix} x_1^{\text{opt}} & 0 & 0 & \ldots 0 \\ 0 & x_2^{\text{opt}} & 0 & \vdots \\ 0 & 0 & x_3^{\text{opt}} & \vdots \\ \vdots & & & \vdots \\ 0 & \ldots & \ldots & x_n^{\text{opt}} \end{bmatrix} \tag{6.3}$$

We shall note that $\boldsymbol{w}\mathbf{L}$ equals $\boldsymbol{x}-\boldsymbol{x}^{\text{opt}}$. Neglecting terms in the Taylor series of order greater than 2, the following expression defining the relative deviation of the function from its optimal value can be obtained:

$$\frac{f(\boldsymbol{x})-f(\boldsymbol{x}^{\text{opt}})}{f(\boldsymbol{x}^{\text{opt}})} = \boldsymbol{w}^T\left[\frac{\mathbf{LGL}}{2f(\boldsymbol{x}^{\text{opt}})}\right]\boldsymbol{w} \tag{6.4}$$

The matrix on the right-hand side of the expression is known as the *square sensitivity matrix*. It is fully defined if $\boldsymbol{x}^{\text{opt}}$, function value and the values of its second derivatives at $\boldsymbol{x}^{\text{opt}}$ are known. The equation (6.4) can thus be used to determine the greatest possible deviation of f from its value at optimum, providing that relative deviations of the variables from the optimum are known. It should be remembered, however, that this expression does not hold if $\boldsymbol{x}^{\text{opt}}$ is not a stationary point of f; such a case is analyzed elsewhere [60].

There may be such instances when, instead of the deviations of the optimum location, parameter deviations should be investigated. Theoretical analysis of this problem and some practical examples can be found in the literature [15, 51, 60].

Occasionally, a reverse problem may be of interest, i.e. what deviations of the coordinates from their optimal values can be allowed without causing the function value to deviate from its value at the optimum by more than a certain value Δf. Again if we assume that $\boldsymbol{x}^{\text{opt}}$ is a stationary point of f, the Taylor series expansion can be used to formulate the following relation:

$$|\Delta f| = \frac{1}{2}\rho^2 \sum_{j=1}^{n}\sum_{k=1}^{n}\left|\frac{\partial^2 f(\boldsymbol{x}^{\text{opt}})}{\partial x_j\,\partial x_k}\right| \tag{6.5}$$

where: ρ – the largest of all deviations of decision variables from their optimal values. Upon transformation of (6.5), we find:

$$\rho = \sqrt{\frac{|2\Delta f|}{\sum_{j=1}^{n}\sum_{k=1}^{n}\left|\frac{\partial^2(\boldsymbol{x}^{\text{opt}})}{\partial x_j\,\partial x_k}\right|}} \tag{6.6}$$

Example 6.1

Consider the function of one variable

$$f(x) = x^2 - 2x$$

which has its minimum at $x^{\text{opt}} = 1$. How much could x deviate from its optimal value without causing the function to increase by more than 1%?

As function value at minimum is -1, the absolute value of its allowable deviation is 0.01.

For a function of one variable, equation (6.6) assumes the form

$$\rho = \sqrt{\frac{2|\Delta f|}{\left|\frac{\partial^2 f(x^{\text{opt}})}{\partial x^2}\right|}}$$

As $\frac{\partial^2 f}{\partial x^2} = 2$, the allowable deviation of the value of the variable is

$$\rho = \sqrt{\frac{2\times 0.01}{2}} = 0.1$$

This means that the function value does not change more than by 1%, providing that the corresponding change in the value of the variable is less than 10%.

6.3 DISCRETE VARIABLES VS. SENSITIVITY ANALYSIS

As already mentioned in Section 4.2, one of the possible approaches to a solution of an optimization problem with discrete variables is to embed it in a continuous problem. Once the continuous problem has been solved, sensitivity analysis can be utilized to find the solution of the original problem. In Fig. 6.1, which gives an example of a function of two variables

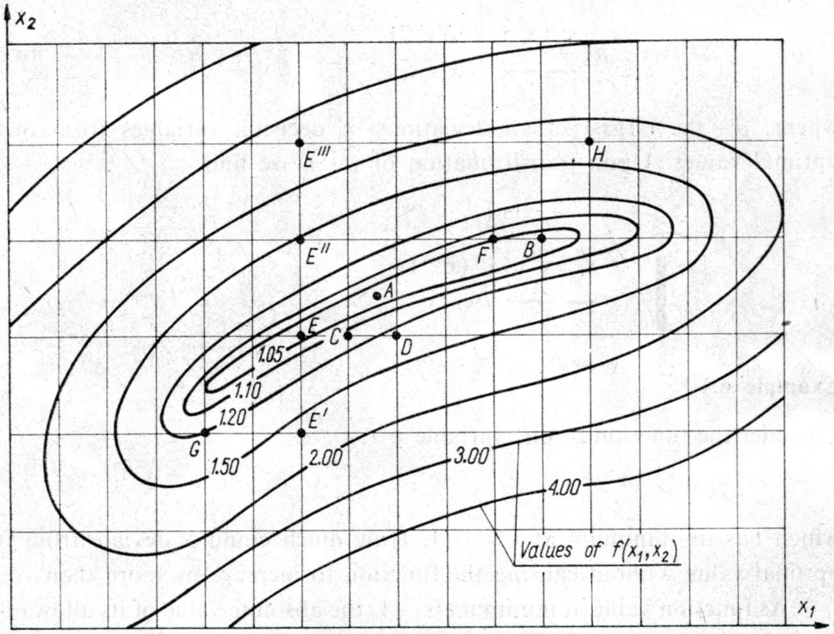

Fig. 6.1 A discrete (at grid points) minimization problem embedded in a continuous problem (ref. [56]). *A* – continuous minimum, *B* – discrete minimum, *C* – the best (in sense of objective function) point neighbouring continuous minimum, *D* – the point closest to continuous minimum. Other points are explained in text.

the difficulties related to the embedding of the discrete problem in a continuous one are illustrated. If both variables are treated as continuous ones, then the minimum of the function can be easily found, even by search in the direction of coordinates (point A).

It can be easily established, however, that the same method will fail in the discrete case (discrete minimum at point B). If an attempt is made to identify the discrete minimum by checking the points neighbouring the continuous minimum, then the procedure must be applied to a large number of points. As it happens, the search is not made easier by the fact that the best (with respect to function value) of points directly neighbouring the continuous minimum, that is point C, is much worse (in the same sense) than other points positioned further away from the continuous minimum, like points E and F.

The procedure for finding the discrete minimum on the basis of sensitivity analysis is as follows:

(1) Locate a discrete point which lies nearest to the continuous minimum (point D in Fig. 6.1).

(2) Starting from this point, find the discrete minimum in the direction of that variable with respect to which the function is least sensitive (in Fig. 6.1, it is x_1, minimum in the direction of x_1 lies at point E).

(3) Starting from the point just found, search the set of feasible solutions in the direction of the second variable x_2; from every intermediate point, determine the discrete minimum in the direction of the first variable (the intermediate points are E′, E″, E‴ and the respective minimum points are G, B, H).

In the example shown in Fig. 6.1, this procedure enables us to find the true discrete minimum B. It should be emphasized that the sequence in which the coordinates are searched is of primary importance. It can be easily proved that a reversed sequence may cause the search to terminate in a point different from B.

This procedure can be generalized to apply to an n-dimensional problem featuring k discrete variables $x_{d1}, x_{d2}, \ldots, x_{dk}$ and $l = n-k$ continuous variables $x_{c1}, x_{c2}, \ldots, x_{cl}$. In a special case, $l = 0$, i.e. all variables are discrete ones.

Let us define the sensitivity index as follows:

$$s_i = \frac{f^*(x_{di}) - f^*(x_{di} + \Delta x_{di})}{\Delta x_{di}} \tag{6.7}$$

where: $f^*(x_{di}) = f(x_{d1}, x_{d2}, \ldots, x_{di}, \ldots, x_{dk}, \min f(x_c))$

$$f^*(x_{di} + \Delta x_{di}) = f(x_{d1}, x_{d2}, \ldots, x_{di} + \Delta x_{di}, \ldots, x_{dk}, \min f(x_c))$$

while $\min f(x_c)$ denotes the values of continuous variables corresponding to the minimum of the function at fixed values of discrete variables; this minimum must be found with the aid of a method suited to problems with continuous variables. Before the search is initiated, the discrete variables should be arranged into a sequence such that

$$s_1 \geqslant s_2 \geqslant \ldots s_i \geqslant s_{i+1} \geqslant \ldots s_k \tag{6.8}$$

Note that s_i values are calculated at the point where the search is initiated, i.e. at the discrete point nearest to the continuous minimum. The search begins in the direction of x_1. Once the point of minimum in that direction has been located, new values of sensitivity indices are calculated. If their sequence is different from that at the starting point, then the search is re-initiated in accordance with the new sequence of variables. If the sequence remains unchanged, then the search continues in the direction of the next variable. At each point checked, the relevant function value is that at the point of minimum in the direction of the previous variable. Once the point of minimum in the direction of the current variable has been found, sensitivity indices are calculated again, and so forth.

The scheme described can be adapted to constrained problems, providing that penalty function methods are applied to their solution. The properties of this scheme and the examples of its application can be found in [56].

CHAPTER 7

Characteristic features of optimization problems in process equipment design

7.1 MATHEMATICAL MODELS AND DESIGN EVALUATION

In the review of equipment design problems given in Chapter 1, a considerable variety of technological and economic problems involving various operations were described. On the formal side, it can be concluded that the quantities considered, i.e. decision variables and model parameters, can be regarded as real variables.

Most equipment design problems can be formulated using deterministic mathematical models. The most common deviation from the determinism is incomplete information on model parameters. This applies, in particular, to the parameters reflecting economic quantities whose values have to be forecast over a long period of operation. Usually, incomplete information is compensated for by some simplifying assumptions.

A characteristic feature of mathematical models appearing in process equipment optimization is the presence of discrete decision variables. This results from the application of various standardized elements, like tubes, plates, flanges, fittings, etc. Another source of discrete variables is in simplification or the reduction of sets of possible types and sizes of the equipment. A characteristic problem in discrete variables is the problem of selection, for given operating conditions, of the most suitable size from an existing set of different sizes.

A special decision problem is the selection of sizes in a set that is a problem concerning simultaneous proportioning and planning of a series of equipment sizes [78]. The optimization of such a series, however, requires mathematical models different from those presented in Chapter 3. Suitable extended optimization models are presented in Section 7.3.

Reflecting a variety of practical problems, the meaning of evaluation criteria used in equipment optimization may vary broadly. Formally, as deterministic mathematical models predominate, the most frequent mathematical form of the evaluation criterion is an objective function. The properties of objective functions encountered in process equipment design have much in common. As a rule, the functions are nonlinear, although it could in some instances be possible to apply a transformation of the variables to obtain linear forms. Along with functions which can be expressed in a relatively simple analytical form, there are also functions which cannot altogether be expressed analytically. This applies, in particular, to process equipment for mass and heat transfer operations because their mathematical models may include transcendental equations. It is very difficult in such cases to recognize important function properties like continuity, differentiability, convexity or unimodality. These properties can be investigated by numerical experiments only.

7.2 PROBLEM OF A SINGLE EQUIPMENT SIZE

The optimal sizing of a single piece of equipment destined for a definite unit operation can be reduced to a mathematical problem of the form consistent with scheme (4.1). Owing to characteristic features of mathematical models and objective functions, it is usually a problem of nonlinear programming

$$\min_{x} f(x)$$

$$\text{subject to} \quad g_i(x) = 0 \qquad i = 1, 2, \ldots p$$

$$h_j(x) \leqslant 0 \qquad j = 1, 2, \ldots q$$

Using various methods of simplification, problems of this kind can be usually reduced to a manageable size. In problems which can be regarded as sensibly formulated, that is where unimportant technological details are omitted, the number of decision variables should not exceed 10, being usually around 2 or 3 and the number of constraints not greater than, say, 4 or 5. This estimate of the average size of mathematical problems, however, is only loosely related to the average level of complexity of design problems. One should realize that the average problem size corresponds to problems solved up to now; it is limited by practical considerations (possibility of quick solution), technical factors (computing equipment available) and psychological factors (possibility of mastering the problem by relying on intuition or on geometrical interpretation). Taking advantage of accumulated experience and applying well-proven mathematical methods and modern computing equipment, it is now possible to undertake and to solve larger problems. The number of decision variables may be up to several dozen and the number of constraints, depending on how complicated the functions involved are, may be of a similar number. Such problems are not unusual, for example, in aerospace or aviation structure design.

7.3 SERIES PROBLEM

Up to now, we have considered design optimization of a single equipment size. This is a part of the cycle of technological activities, aimed at satisfying the needs of a certain user, or of a group of users, by a single design. A scheme of this cycle is shown in Fig. 7.1. The essence of optimization procedures is to forecast – at least partially – the effects associated with manufacturing, distribution and operation of equipment, and to find a design decision that is most suitable with respect to these effects.

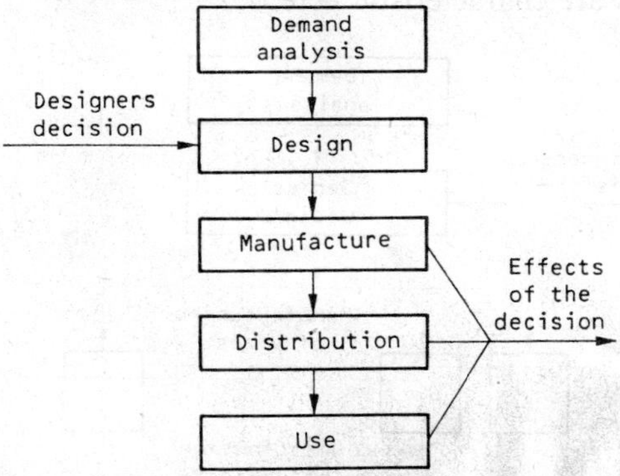

Fig. 7.1 A cycle of technological activities associated with market introduction of a single equipment design.

The problem of a series design arises when, for a certain equipment type, a group of prospective users ex'sts whose needs, however, differ qualitatively. It may be technologically impracticable or economically unfeasible to satisfy these needs by a single equipment design [38]. The design problem is then concerned with a series, that is a set of different equipment sizes. As practical experience proves, the sensible design of an equipment series is especially important in the matter of satisfying conflicting interests of manufactures and users of the equipment, while also contributing to the improvement of the design of plants where the equipment is used. The conflict of interests arises from the fact that the manufacturers wish to keep the costs down by reducing the number of sizes, while the users prefer a large variety from which to choose. A well-designed series is a result of a trade-in decision that ensures the largest overall benefits. At the same time, the ordered variety of equipment sizes enables plant designers to improve process design without conflicting with economical equipment production.

The economic gains associated with the availability of a series arise from the repeated use of identical equipment sizes in a number of various plants. The quality of the series has thus to be evaluated on the basis of economic effects of operating all equipment units in the relevant plants. When designing the series, these economic effects are forecast and engineering decisions are sought to maximize the future utilization.

The economic effects associated with a series design can be forecast following the observation that if a set of various equipment sizes exists, then a change occurs in the cycle of technological activities associated with satisfying user's needs. The relevant cycle is schematically shown in Fig. 7.2. Two factors are characteristic here:

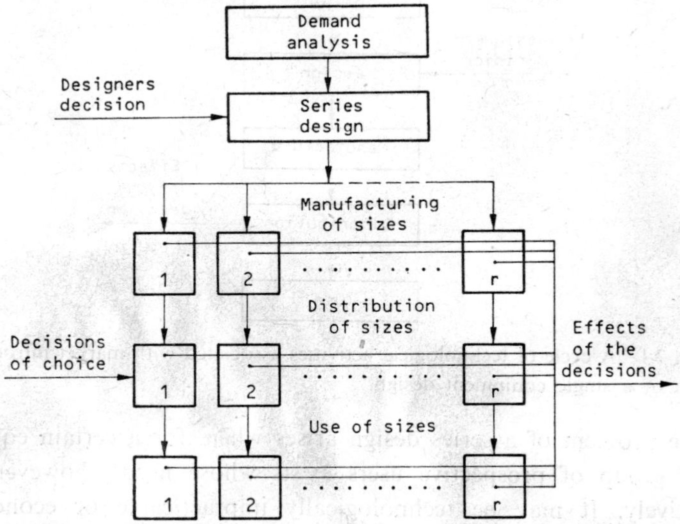

Fig. 7.2 A cycle of technological activities associated with market introduction of a type series.

(1) The consequences of the designer's decision appear separately for every equipment size in the series.

(2) The choices of individual equipment sizes made by various users, must be taken into account as additional decisions contributing to the final effect of the decision taken by the series' designer.

In the mathematical decision model describing series design [78], the basic concept is a design or specification of one equipment size (design of a unit or the unit itself), considered as point x in multidimensional

Euclidean space R^m. (This convention is, in fact, identical with that used in classical design optimization considered in Chapters 2 and 3.) It is generally enough if, in this space, points satisfying obvious physical, standardization or other conditions are considered. The reasoning is then limited to a closed subspace X.

Each user may operate the equipment in his individual manner (mode of operation) which can be considered as point $\boldsymbol{y}$ in another multidimensional Euclidean space R^m. Here, the coordinate system is related to process quantities which are important both in design and use. These quantities may be chosen very flexibly, for example, if the equipment is operated at variable pressure, then both minimum and maximum pressures qualify as process variables. The designer has to consider only the points in R^m which satisfy obvious physical and technological conditions (for example, temperature and pressure limits associated with the equipment type). The reasoning is thus concerned with certain closed subspace Y which will hereafter be called operation space.

A mathematical model of a single equipment size is created according to the principles stated in Chapter 3. Let us assume the most frequent case characterized by a system of constraints

$$g_i(\boldsymbol{x}, \boldsymbol{y}) = 0 \qquad i = 1, 2, \ldots p$$
$$h_i(\boldsymbol{x}, \boldsymbol{y}) \leqslant 0 \qquad j = 1, 2, \ldots q$$

Such a constraint system defines two functions which link together designs and modes of operation, Fig. 7.3:

(1) ρ – assigning elements of space $\Omega(Y)$ to the elements of space X, where $\Omega(Y)$ is a set of closed subsets of Y. A set of allowable modes of operation $\rho(\boldsymbol{x})$ is assigned to each design $\boldsymbol{x}$ ($\rho(\boldsymbol{x})$ is a set of all $\boldsymbol{y}$'s that satisfy, together with given $\boldsymbol{x}$, the constraint system).

(2) η – assigning elements of space $\Omega(X)$ to the elements of space Y,

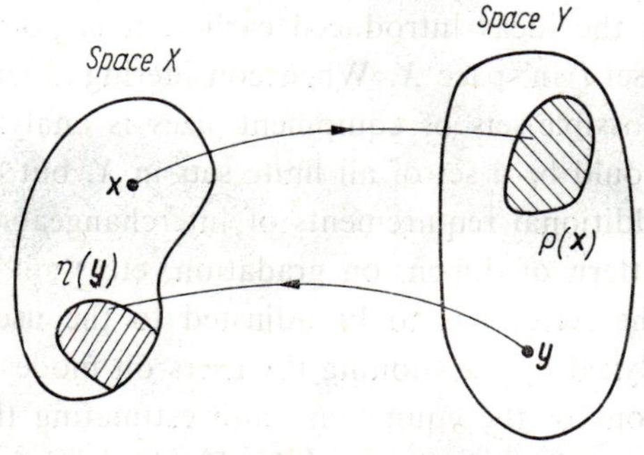

Fig. 7.3 Relationships between designs $\boldsymbol{x}$ and modes of operation $\boldsymbol{y}$, as defined by the constraint system.

where $\Omega(X)$ is a set of closed subsets of X. A set of feasible designs $\eta(\boldsymbol{y})$ is assigned to each mode of operation $\boldsymbol{y}$ ($\eta(\boldsymbol{y})$ is a set of all $\boldsymbol{x}$'s that satisfy, together with given $\boldsymbol{y}$, the constraint system).

When evaluating a design from the points of view of different users, it is necessary to account for variable operation modes. The objective function expressing the evaluation criterion must then be defined on the Cartesian product $X \times Y$. As an example, a function expressing total or annual cost of investment and operation can be named. It is evident that for almost all types of process equipment, this function depends on design and process quantities.

Let us denote such an objective function by $d(\boldsymbol{x}, \boldsymbol{y})$ and assume that it should be minimized. Two elementary optimization problems are then possible.

1. Optimization of the design for a given operation mode $\boldsymbol{y}^*$, which means searching for such $\boldsymbol{x}^{*0}$ that

$$d(\boldsymbol{x}^{*0}, \boldsymbol{y}^*) = \min_{\boldsymbol{x} \in \eta(\boldsymbol{y}^*)} d(\boldsymbol{x}, \boldsymbol{y}^*)$$

The notation reminds us that the solution must satisfy the constraint system – the set of allowable solutions is $\eta(\boldsymbol{y}^*)$.

2. Optimization of the operation mode for given equipment design $\boldsymbol{x}'$, which means searching for such $\boldsymbol{y}'^0$ that

$$d(\boldsymbol{x}', \boldsymbol{y}'^0) = \min_{\boldsymbol{y} \in \rho(\boldsymbol{x}')} d(\boldsymbol{x}', \boldsymbol{y})$$

Again, the notation reminds us that the constraint system must be satisfied – the set of allowable modes of operation is $\rho(\boldsymbol{x}')$.

Both situations described above appear in the design of a series. The first scheme indicates the user's dilemma of choosing the size which fits best to actual needs and the second one takes the form of asking how to use most efficiently the proposed size.

Returning to the ideas introduced earlier, it is possible to interpret a series as a finite set Γ in space X. When considering different series, a certain space W of possible sets of equipment sizes is analyzed. In the most general case W would be a set of all finite sets in X, but usually these sets have to satisfy additional requirements of interchangeability of parts and subassemblies, pattern of dimension gradation, etc.

Obviously, the series has to be adjusted to the user's needs. These needs can be analyzed by questioning the users on modes of operation for planned applications of the equipment, and estimating the probability of different modes of operation in space Y. If obtained from a sufficiently large number of users, such information places an array of points in space

Y. This enables a probability measure μ to be defined on the subsets of space Y. In a special case, probability density could also be defined: $d\mu(y) = \psi(y)dy$, where ψ is the density function defined on the elements of space Y.

When a set of equipment sizes is produced, it may become necessary to modify the principles of evaluation of a single design. Firstly, the partition of total demand among different sizes affects the production volume of each size and this in turn has its bearing upon manufacturing cost per unit. Secondly, certain features of the series may result in interchangeability of parts or subassemblies, different degrees of interchangeability may also lead to different costs per unit. Thus, any single design should be evaluated against the background of the entire series. As an example, let us consider design $\boldsymbol{x}$ and two series Γ_1 and Γ_2 such that $\boldsymbol{x}$ is a type dimension in the first case only, i.e. $\boldsymbol{x} \in \Gamma_1$ but $\boldsymbol{x} \notin \Gamma_2$. Characteristically, the cost of $\boldsymbol{x}$ will be lower in the first case, because a well-prepared production of a type dimension selected in advance is cheaper than manufacturing one size according to a unique design.

If we are thus willing to account simultaneously for the interests of equipment manufacturer and of users, then the objective function for equipment design evaluation should be defined on set $X \times Y \times W$. More accurately, it can be defined on a set of such triples $(\boldsymbol{x}, \boldsymbol{y}, \Gamma)$ in which $\boldsymbol{y} \in \rho(\boldsymbol{x})$, because the design can be evaluated only if the mode of operation is physically realizable. However, the function can be conveniently expanded on the entire product $X \times Y \times W$, by assuming that its value is positive and infinite for $\boldsymbol{y} \notin \rho(\boldsymbol{x})$. Such a function will hereafter be denoted f.

If a series is introduced, then different equipment sizes are produced by one or a few manufacturers and operated by numerous users. The objective function explained above makes it possible to express the economic consequences of series production per single unit produced. The overall effect can be measured by mean weighted value of the function in space Y, the weight being the probability of demand. Let us consider the series as a finite set of equipment sizes $\Gamma = \{\boldsymbol{x}_i; i = 1, 2, \ldots\}$. The only initial assumption is that different sizes are used in disjoint sets of modes of operation $\Lambda_i; i = 1, 2, \ldots$ (operation modes must be feasible, which means that $\Lambda_i \subset \rho(\boldsymbol{x}_i)$). Using the idea that has been adopted earlier among others, in the optimization of parametric series† and safety standards‡,

† Sittig, J., Economically Optimal Standardization *Proc. 1st Symp. Applications of Mathematics to Standardization,* Warsaw, 1969.

‡ Lind, N. C., Basler, F., Safety Level Decisions, *Proc. Int. Conf. on Planning and Design of Tall Buildings,* Bethlehem PA, 1972.

the overall effect of applications of all equipment sizes can be expressed as

$$F(\Gamma) = \sum_i \int_{\Lambda_i} f(x_i, y, \Gamma)\, d\mu(y) \tag{7.1}$$

The integral appearing in this expression should be interpreted as a Lebesgue–Stieltjes integral, so that the definition of $F(\Gamma)$ covers a wide class of cases featuring different properties of space Y.

The sets of modes of operation assigned to individual equipment sizes are defined by the user's decisions to choose their sizes from the series. It can be generally assumed that these decisions follow the trend towards optimum equipment utilization and thus, for every mode of operation, a size is chosen according to the scheme explained above. As a result, the objective function $f(x, y, \Gamma)$ always takes the lowest value with respect to $x \in \Gamma$, which can be described in the equivalent form of a new function defined on set $Y \times W$

$$f_\Gamma(y, \Gamma) = \min_{x \in \Gamma} f(x, y, \Gamma) \tag{7.2}$$

Any size $x \in \Gamma$ should be used only if it minimizes the objective function. The definition of a set of rational modes of operation $\Phi_\Gamma(x, \Gamma)$ follows herefrom, viz., this is a set of such y's that $f(x, y, \Gamma) = f_\Gamma(y, \Gamma)$.

Suppose a compulsory standard is issued defining a series to which the users must confine their choice of equipment sizes. The series should give all the users the possibility of finding a suitable size. Mathematically, this can be interpreted as a requirement that the sets of feasible modes of operation form a finite covering of space Y (cf. Appendix 1)

$$Y = \bigcup_{x \in \Gamma} \rho(x) \tag{7.3}$$

The overall effect of series applications, providing the series is optimally utilized, can be expressed as

$$G(\Gamma) = \int_Y f_\Gamma(y, \Gamma)\, d\mu(y) \tag{7.4}$$

Recalling that $\Gamma = \{x_i; i = 1, 2, \ldots\}$, this is equivalent to

$$G(\Gamma) = \sum_{i=1}^{r} \int_{\Phi_i} f(x_i, y, \Gamma)\, d\mu(y) \tag{7.5}$$

where $\quad \Phi_i = \Phi_\Gamma(x_i, \Gamma)$

Optimal series should minimize functional G over space W while satisfying the constraint which requires, according to Eq. (7.3), a finite covering of space Y to exist.

There is no general theory of mathematical problems of this kind. However, many practical cases can be successfully solved using conventional optimization techniques. As a rule, a two-level approach can be adopted, which means that for different r values, optimal sets $\{x_i; i = 1, 2, \ldots r\}$ are found. Finally, the global solution is identified as

$$\min_{r} \min_{\{x_i\}} \sum_{i=1}^{r} \int_{\Phi_i} f(y, x_1, x_2 \ldots x_r)\, d\mu(y)$$

It should be added that instead of (7.3), an equivalent constraint formulation can be conveniently introduced using the idea of sets of rational modes of operation

$$Y = \bigcup_{i} \Phi_i$$

Examples of series determined with the aid of the theory presented above are given in Chapters 10 and 11.

CHAPTER 8

Mechanical elements

8.1 GENERAL PURPOSE ELEMENTS

As with other kinds of machinery, process equipment may include general purpose mechanical elements such as beams, shafts, bearings, rotating discs, gears, transmissions, simple mechanisms, etc. As such elements are being widely applied, optimization of their design has been pursued by mechanical engineers for many years. Numerous solutions of optimization problems can be found in the literature, concerning mechnical design under various evaluation criteria and various constraints [22, 30, 31, 62, 72, 73]. Using and refining them is now almost a standard routine in general mechanical design and it can also be implemented in process equipment design.

Characteristic features of the majority of problems concerning optimization of elements are as follows:

– decision variables correspond to the physical properties of the part designed (most often – dimensions),

– parameters correspond to material properties (allowable stress, elasticity coefficient, etc.), characteristic load figures (forces, moments, etc.) and characteristic design requirement indices (allowable deflection, allowable angle of torsion, etc.),

– mathematical models consist mainly of geometrical relations, mechanics of materials formulae, technological requirements, etc.). The models are mainly deterministic and thus design evaluation can be based on objective functions expressing weight, various structural indices (for example, characteristic parameters of stress distribution, compactness ratio, etc.) or operational indices (e.g. for bearings: power loss, wear, etc.).

A classical example of the optimization of a mechanical element is the problem of flexure of a cantilever beam of minimum weight. The solutions to this problem formulated for various beam cross-sections and various load conditions can be found in the literature [8, 22, 62].

Another classic is the following example of the optimization of dimensions of a shaft subject to bending and torsion.

Example 8.1

Diameter d and length l of a cantilever shaft of least weight is to be found. Shaft cross-section is circular, and the loads are: bending moment and torsion moment characterized by maximal instantaneous values M_b and M_t, respectively (Fig. 8.1). Shaft material is characterized by: allowable

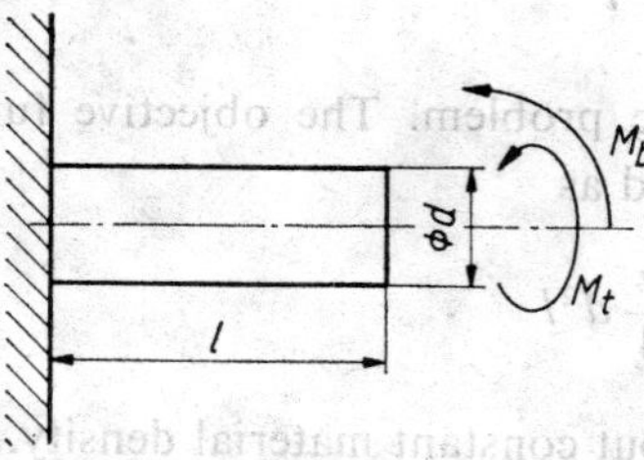

Fig. 8.1 Cantilever shaft subject to bending and torsion.

stress k, Young's modulus E and modulus of elasticity G. Diameter and length must range within d_{min} to d_{max} and l_{min} to l_{max}, respectively. Angle of torsion must not exceed allowable value θ, and maximum allowable deflection is t.

Decision variables in this problem are:

– diameter d, where

$$d_{min} \leqslant d \leqslant d_{max}$$

– length l, where

$$l_{min} \leqslant l \leqslant l_{max}$$

Decision space is a bounded region in a two-dimensional space (a rectangle in a plane).

Parameters in this problem are:

– material constants k, E, G;

– characteristic load figures M_b, M_t;

– characteristic design requirement indices t, θ.

Parameter space is seven-dimensional.

The mathematical model is set up on the basis of relations describing the mechanics of materials phenomena and design requirements:

– allowable equivalent stress, according to the hypothesis of energy of deformation

$$\sqrt{M_b^2 + \frac{3}{4} M_t^2} \frac{32}{\pi d^3} \leqslant k \tag{8.1}$$

– allowable angle of torsion

$$\frac{M_t}{G} \frac{32l}{\pi d^4} \leqslant \theta \tag{8.2}$$

– allowable deflection

$$M_b \frac{32l^2}{E\pi d^4} \leqslant t \tag{8.3}$$

This is a deterministic problem. The objective function expressing shaft mass can be formulated as

$$f(d, l) = \rho \frac{\pi}{4} d^2 l \tag{8.4}$$

where: ρ – unknown but constant material density. Formula (8.4) can be simplified by ignoring the constants so that

$$f^*(d, l) = d^2 l \tag{8.5}$$

The problem composed of constraints (8.1)–(8.3) and objective function (8.5) is nonlinear with respect to variables d and l. However, it can be transformed into a linear problem by taking the logarithms of all expressions and by substituting new variables

$$\ln d = x_1$$

$$\ln l = x_2$$

The result of this transformation is that the expressions containing products or quotients of original variables are now substituted by expressions containing sums or differences of new variables. Upon adjusting the transformed expressions, the constraints assume the form

$$x_1 \geqslant \frac{1}{3} \ln \left(\sqrt{M_b^2 + \frac{3}{4} M_t^2} \frac{32}{\pi k} \right) \tag{8.6}$$

$$4x_1 - x_2 \geqslant \ln \left(\frac{M_t}{G} \frac{32}{\pi\theta} \right) \tag{8.7}$$

$$2x_1 - x_2 \geqslant \frac{1}{2} \ln \left(\frac{32 M_b}{E\pi t} \right) \tag{8.8}$$

The problem includes also the constraints imposed on the values of variables

$$\ln d_{\min} \leqslant x_1 \leqslant \ln d_{\max} \tag{8.9}$$

$$\ln l_{\min} \leqslant x_2 \leqslant \ln l_{\max} \tag{8.10}$$

The transformed objective function is

$$f^m(x_1, x_2) = 2x_1 + x_2 \tag{8.11}$$

The problem of minimization of function (8.11) subject to constraints (8.6)–(8.10) can be solved graphically (Fig. 8.2). It follows from the graph that constraint (8.6) is redundant.

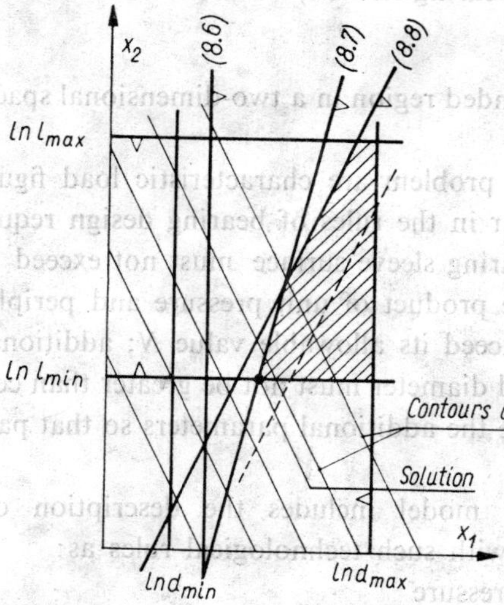

Fig. 8.2 Graphical solution to the shaft optimization problem. Dashed line represents the position of constraint (8.8) in case of very small allowable deflection (see discussion in the text).

It is worth noting that the location of the solution point at the intercept of lines of least length and of allowable angle of torsion may be drastically changed if problem parameters are changed. For example, reducing allowable deflection t to a sufficiently small value may cause the line representing constraint (8.8) to move to the right (dashed line) which wolud be accompanied by moving the solution to the least-length line intercept. Then, constraint (8.7) would be redundant.

The engineering meaning of the optimization of mechanical elements is not necessarily always as obvious as in the example presented. In problem formulation, difficulties may arise concerning the objective function, with respect to either its physical meaning or the mathematical formulation.

Example 8.2

Determine the evaluation criterion and find the optimal design of a radial slide bearing with a self-lubricating porous material bearing sleeve. Bearing capacity is R and shaft rotational speed is n [rpm]. Shaft neck diameter and active length of the bearing sleeve must not exceed allowable values d_{max} and l_{max}, respectively.

Decision variables in this problem are:

– shaft neck diameter d, where

$$d_{min} \leqslant d \leqslant d_{max}$$

– active length of bearing sleeve l, where

$$0 < l \leqslant l_{max}$$

Decision space is a bounded region in a two-dimensional space (a rectangle in a plane).

Parameters in this problem are characteristic load figures R and n; other parameters appear in the rules of bearing design requiring that the unit pressure at the bearing sleeve surface must not exceed the allowable value p_{max} and that the product of unit pressure and peripheral speed of shaft neck must not exceed its allowable value N; additionally, the ratio of shaft neck length and diameter must not be greater than certain value L. Thus, N, p_{max} and L are the additional parameters so that parameter space is five-dimensional.

The mathematical model includes the description of mechanical phenomena combined with such technological rules as:

– allowable unit pressure

$$\frac{R}{dl} \leqslant p_{max}$$

– allowable product of unit pressure and peripheral speed

$$\frac{R}{dl}\frac{\pi n}{60}d \leqslant N$$

A geometrical constraint must also be accounted for:

$$\frac{l}{d} \leqslant L$$

This is a deterministic problem formulation. The choice of an objective function is by no means evident. For example, the weight of shaft neck cannot be used in the evaluation of solutions as the shape and weight of the remaining part of the shaft are unknown. The expected bearing life would be more justified as a criterion, but treating it literally as an objective function would require searching for additional design data. This difficulty can be resolved, however, if we focus our interest on the fact that the expected life increases with decreasing product of unit pressure and peripheral velocity of the neck. The relevant objective function can thus be formulated as

$$f^*(l) = \frac{R\pi n}{60} \frac{1}{l}$$

and after ignoring the constants

$$f(l) = \frac{1}{l}$$

The requirement that the maximum expected life should be attained is thus equivalent to the minimization of the above function. In other words, shaft neck should be of maximum possible length.

A graphical presentation of the problem of optimization of bearing dimensions is shown in Fig. 8.3. Two different solutions are indicated.

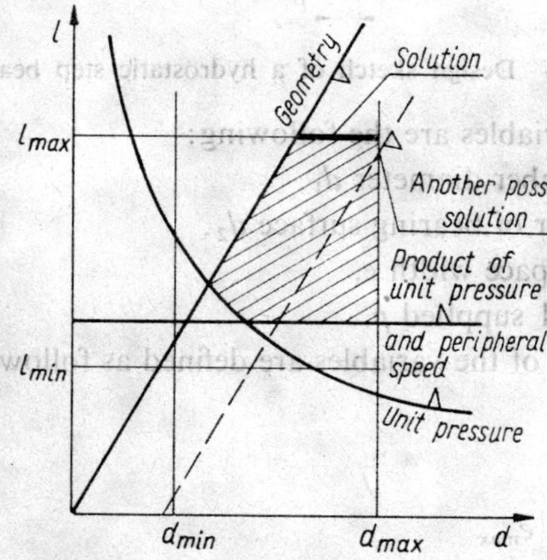

Fig. 8.3 Graphical solution to the bearing optimization problem. Dashed line represents the position of the geometrical constraint at reduced value of parameter L (see discussion in the text).

Firstly, many points of equal quality are possible on the line of maximum allowable length. Secondly, another solution can be obtained at smaller values of length-to-diameter ratio L: assuming the geometrical constraint is shifted into the position indicated by the dashed line, the solution is defined by the intersection point with the allowable diameter d_{max} line.

The examples presented above are exceptionally simple as they have been chosen to illustrate the general idea od solving problems of this kind. In most problems of optimization of mechanical elements, the mathematical models are much more complicated and, in addition to that, it may be impossible in multidimensional decision spaces to recognize the properties of the solution quickly.

Example 8.3

The evaluation criterion is to be determined and the most advantageous design found for a hydrostatic step bearing loaded by force R at a shaft rotational speed n. The design scheme of the bearing is shown in Fig. 8.4.

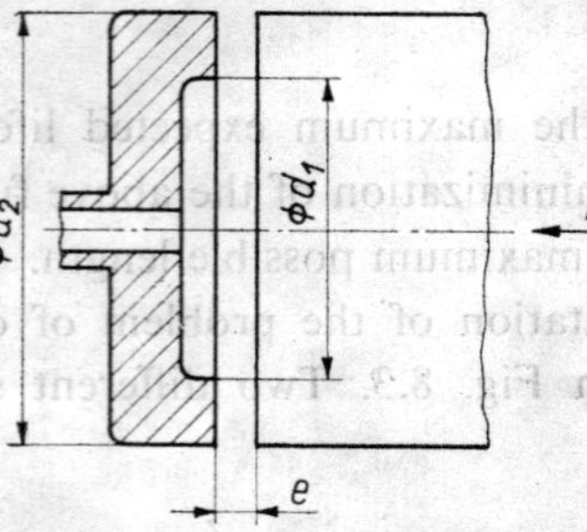

Fig. 8.4 Design sketch of a hydrostatic step bearing.

The decision variables are the following:
– pressure chamber diameter d_1,
– outer diameter of bearing surface d_2,
– bearing interspace width e,
– pressure of oil supplied p.

The allowable values of the variables are defined as follows:

$$d_1 \geqslant d_{1\min} \tag{8.12}$$

$$d_2 \leqslant d_{2\max} \tag{8.13}$$

$$e_{\min} \leqslant e \leqslant e_{\max} \tag{8.14}$$

While it is evident that characteristic load figures R and n belong to the parameters, other parameters must be identified in the course of formulation of mathematical model and objective function.

A geometrical constraint must be included in the model:

$$d_1 < d_2 \tag{8.15}$$

Unit pressure at the bearing surface must not exceed its allowable value

$$\frac{4R}{\pi(d_1^2 - d_2^2)} \leqslant p_{\max} \tag{8.16}$$

Essentially, the design of the step bearing consists of proportioning the bearing surface in such a way as to minimize the power loss. The power loss is a sum of power required for oil pumping and friction power. The power consumed in the oil pump is

$$P_p = \frac{pV}{\eta}$$

where V – volume flow of oil, η – efficiency of the pump.

The oil flow between the pump and the pressure chamber of the bearing is accompanied by a pressure drop the consequences of which can be approximately expressed as

$$p_0 = ap$$

where: p_0 – oil pressure in the chamber, a – known constant less than 1.

According to the Poiseuille equation, the volume flow depends on oil pressure in the chamber and on the free area of bearing interspace through which oil flows out

$$V = p_0 \pi d_m e \frac{e^2}{12\mu \frac{d_2 - d_1}{2}}$$

where: $d_m = \frac{d_1 + d_2}{2}$, μ – absolute oil viscosity.

The pressure in the chamber must ensure balancing of the bearing load

$$p_0 = \frac{R}{\frac{\pi d_m^2}{4}}$$

Accordingly, pumping effect can be expressed as

$$P_p = \frac{p}{\eta} \frac{R}{\frac{\pi d_m^2}{4}} \pi d_m e^3 \frac{1}{12\mu \frac{d_2 - d_1}{2}} = \frac{R}{3\eta\mu} \frac{pe^3}{d_2^2 - d_1^2}$$

The remainder of the power loss, that is friction power

$$P_f = Fu$$

where: F – friction force, u – peripheral speed (both quantities defined at mean diameter of the bearing surface).

Friction force

$$F = \tau S$$

where: τ – shear stress in oil layer, S – bearing surface area.

Shear stress can be calculated from the Newton formula

$$\tau = \mu \frac{u}{e} = \mu \frac{\omega d_m}{2e}$$

where: $\omega = \dfrac{2\pi n}{60}$ – shaft angular velocity.

Bearing surface area

$$S = \pi d_m \frac{d_2 - d_1}{2}$$

Power loss due to friction

$$P_f = \mu \frac{2\pi n}{60} \frac{d_m}{2e} \pi d_m \frac{d_2 - d_1}{2} \frac{2\pi n}{60} \frac{d_m}{2} = \frac{\mu \pi^3 n^2}{7200} \left(\frac{d_1 + d_2}{2} \right)^3 \frac{d_2 - d_1}{e}$$

It can now be concluded that additional parameters are: allowable unit pressure $p_{\max}$, absolute oil viscosity μ and pumping efficiency η. Parameter space is thus five-dimensional and exactly one point is given in it, as this is a deterministic problem. The relations given above make it possible to formulate the objective function expressing the total power loss in the bearing:

$$f(d_1, d_2, e, p) = P_p + P_f = \frac{R}{3\eta\mu} \frac{pe^3}{d_2^3 - d_1^2} + \frac{\mu \pi^3 n^2}{7200} \frac{d_2 - d_1}{e} \left(\frac{d_1 + d_2}{2} \right)^3 \tag{8.17}$$

The optimization problem consists thus in minimizing function (8.17) subject to constraints (8.15) and (8.16). The objective function is nonlinear, continuous and differentiable and constraint functions are also continuous, which makes it possible to employ various direct methods to the problem solution. Before initiating the computations, it is advisable to simplify the problem form using a transformation of variables according to formulae (4.3)–(4.6). This would enable us to eliminate constraints (8.12) to (8.14).

8.2 ELEMENTS ASSOCIATED WITH HEAT TRANSFER AND FLUID FLOW

Mechanical problems of process equipment design are often coupled with various problems of the design of the process inside the equipment. In the proportioning of mechanical elements, of prime significance are heat transfer and fluid flow effects. The examples of optimal designs related to the evaluation criteria with such phenomena taken into account can be increasingly often found in the literature [85]. Optimization problems of this kind seem to be more diversified than the problems described in the previous section. In particular, it may be sometimes appropriate to consider the design of the process in which the element is involved, rather than the element design itself. Optimal process design can provide information or optimal element design (using, for example, continuity equations, pressure loss equations, etc.).

The properties of problems of this type are as follows:

– decision variables represent physical properties of the equipment unit considered or process quantities;

– parameters represent material constants, physical properties of materials used in the process, process quantities and characteristic coefficients of design requirements;

– mathematical models consist of relations between process quantities, descriptions of the phenomena in the process and design requirements; the models are either deterministic or with incomplete information;

– quality indices are considerably diversified; simple physical quantities (time, flow) can be used, as well as design indices (e.g. weight per unit capacity) or indices related to process quality (e.g. energy consumption per unit capacity).

A classical problem of element optimization with thermal phenomena taken into account is the determination of the most economical thickness of thermal insulation [62]. A number of problems have been formulated on the optimization of fins on heat transfer surfaces [14, 46, 63], although some of them are of theoretical interest only as either the design evaluation criteria applied have nothing to do with practical needs, or the theoretical fin shape can never be practically realized.

Example 8.4 below is based on information taken from the literature [46]. Its formulation is accurate and the results obtained can be useful in practical design.

Example 8.4

Diameter D, thickness 2δ and spacing b of circular fins on a pipe of diameter d are to be determined (Fig. 8.5) for which the best utilization of the fin material, with heat transfer by natural convection, is obtained. The thermal conductivity k_r of the pipe wall and fin material is known. The form of the empirical relation describing heat transfer by natural convection from a cylindrical finned surface is also known to be:

$$Nu = \frac{1}{6\pi}\frac{b}{D}\,Gr\,Pr\left[1-\exp\left(\frac{-25.3D}{bGr\,Pr}\right)^{3/4}\right] \tag{8.18}$$

where: Nu, Gr, Pr – Nusselt, Grashof and Prandtl numbers defined at a fluid temperature equal to that of the fin surface.

The decision variables are as follows:
- fin diameter D,
- fin half-thickness δ,
- fin spacing b.

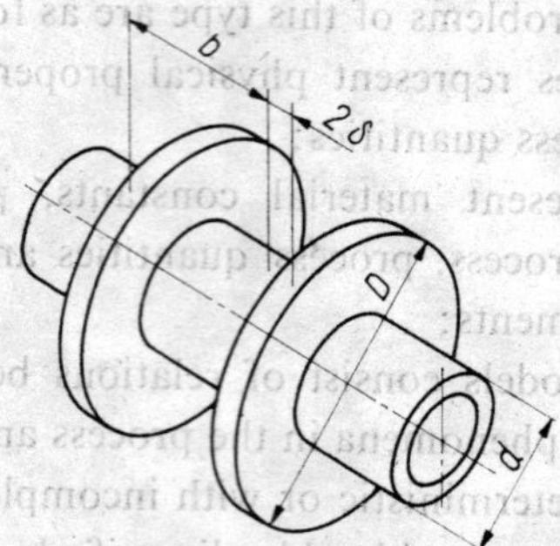

Fig. 8.5 Shape of a finned pipe.

The parameters:
- pipe diameter d,
- thermal conductivity of pipe wall and fin material k_r,
- pipe outer surface temperature t_r,
- fluid temperature at a large distance from heat transfer surface t_f.

The mathematical model is rather simple as it consists of obvious geometrical conditions

$$D > d, \quad b > 2\delta \tag{8.19}$$

In contrast with the model, the objective function is somewhat awkward. The index of utilization of fin material (decreasing when the utilization improves) can be expressed as

$$f(D, \delta, b) = \frac{V_l}{q_l} \tag{8.20}$$

where: V_l – fin material volume, q_l – heat transferred (both quantities reduced to the unit pipe length). In order to transform equation (8.20) into a form suited to an optimization problem, a lengthy series of relations must be considered. Material volume V_l is

$$V_l = \frac{1}{b} \frac{\pi(D^2 - d^2)}{2} \delta \tag{8.21}$$

Heat transferred per unit pipe length

$$q_l = \frac{1}{b} q_z + q_r \tag{8.22}$$

where: q_z – heat transferred from one fin, q_r – heat transferred from the pipe surface.

By introducing equivalent fin diameter

$$D_z = D + 2\delta \tag{8.23}$$

the effect of one fin surface can be expressed by

$$q_z = 2\pi h d(t_r - t_f)\frac{\Lambda}{N} \tag{8.24}$$

where: h – surface film conductance,

$$N = 2\sqrt{\frac{h}{k\delta}} \tag{8.25}$$

$$\Lambda = \frac{I_1(ND_z)K_1(Nd) - I_1(Nd)K_1(ND_z)}{I_0(Nd)K_1(ND_z) + I_1(ND_z)K_0(Nd)} \tag{8.26}$$

and I_n, K_n ($n = 0, 1$) are modified Bessel functions of the first and second kind, of order n.

Formulae (8.24)–(8.26) contain surface film conductance h which depends on temperatures t_r and t_f. This can be calculated using the concepts of average fin surface temperature t_m and fin efficiency η. These are related by

$$t_m - t_f = \eta(t_r - t_f) \tag{8.27}$$

On the other hand, fin efficiency depends on the radial temperature distribution

$$\eta = \frac{16d\Lambda}{N(D_z^2 - d^2)} \tag{8.28}$$

In the calculation of surface film conductance h, definitions of dimensionless numbers are also important

$$Gr = \frac{g\beta(t_r - t_f)D^3}{\nu^2} \tag{8.29}$$

$$Nu = \frac{hD}{k_f} \tag{8.30}$$

where: g – acceleration due to gravity, β – coefficient of volume expansion of fluid, ν – kinematic viscosity of fluid, k_f – thermal conductivity of fluid.

It will be noted that the definition of Prandtl number Pr need not be introduced here, as values for various fluids can be found in the literature.

Upon determination of the Nusselt number from formula (8.18), surface film conductance can be calculated from transformed equation (8.30)

$$h = \frac{Nu\, k_f}{D} \tag{8.31}$$

As the values of k_f needed in formula (8.31) and ν needed in (8.29) should be calculated at temperature t_m which is unknown initially, it is necessary to use successive approximations as shown by the logic diagram in Fig. 8.6. Once h has been determined, it is possible to calculate thermal effect q_z.

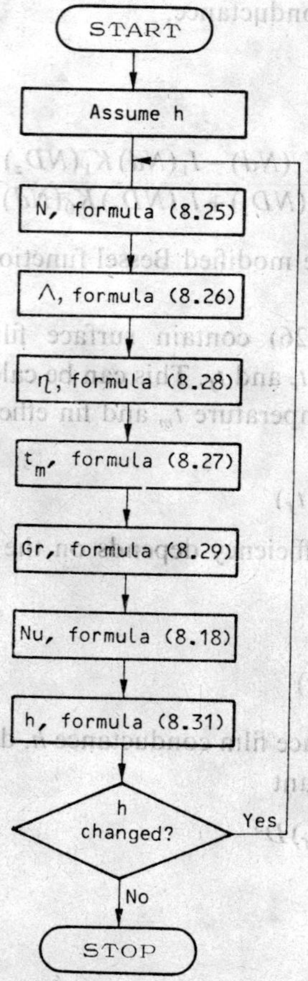

Fig. 8.6 Logic diagram of the calculation of surface film conductance at a finned pipe.

Surface film conductance at the pipe surface between fins can be assumed to be approximately equal to that at the fin surface. The heat transfer from the bare pipe surface is thus

$$q_r = \pi h d\left(1 - \frac{2\delta}{b}\right)(t_r - t_f) \tag{8.32}$$

Using equations (8.18), (8.21)–(8.32) and inserting (8.21) and (8.22) into (8.20), it is possible at known parameter values to determine the value of the objective function for every design decision represented by known values of D, b and δ.

The formulae given above include certain simplifications which may limit the accuracy of calculation of surface film conductance h. The comparison between calculated results and published empirical data [46], however, reveals that the consistency is quite good (Fig. 8.7).

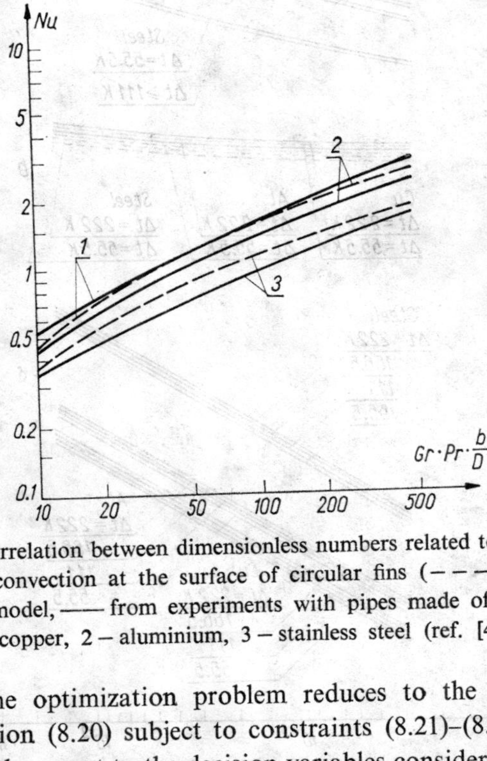

Fig. 8.7 Correlation between dimensionless numbers related to heat transfer by natural convection at the surface of circular fins (– – – according to theoretical model, —— from experiments with pipes made of different materials). 1 – copper, 2 – aluminium, 3 – stainless steel (ref. [46]).

Finally, the optimization problem reduces to the minimization of objective function (8.20) subject to constraints (8.21)–(8.32). The problem is nonlinear with respect to the decision variables considered and the objective function cannot be expressed analytically. The only way to solve the problem is to use direct methods. In addition, as the iterative computations according to the scheme in Fig. 8.6 are always associated with some inaccuracy of results, some scatter of the calculated function values about an ideal theoretical surface cannot be avoided. The method of solution must be insensitive to such a problem.

Before a solution is attempted, it is advisable to simplify the form of the problem by applying variable transformations which make it possible to eliminate constraints (8.19)

$$D = d + e^x, \qquad b = 2\delta + e^y$$

where x and y are new variables. Examples of results taken from [46] are given in Fig. 8.8.

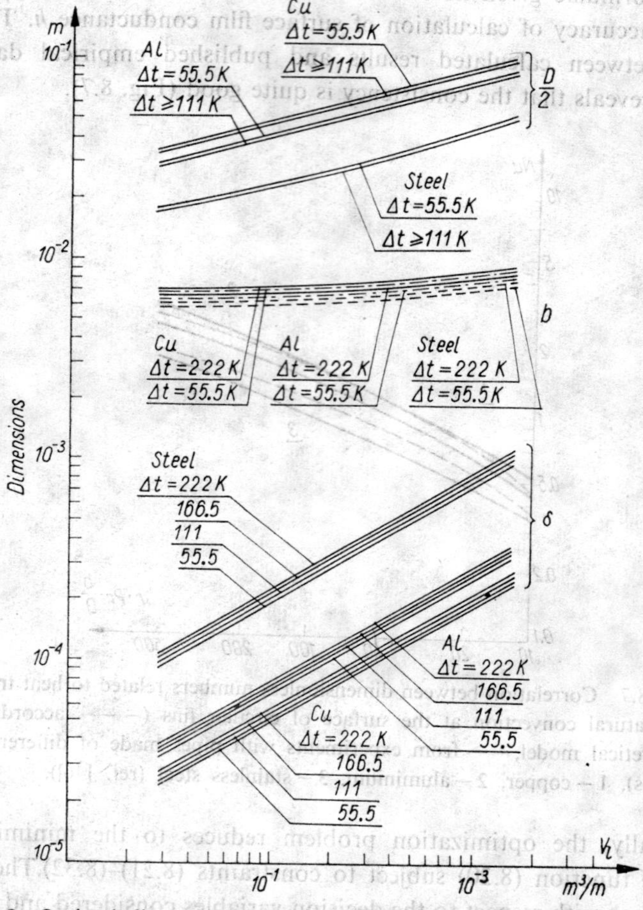

Fig. 8.8 Optimal dimensions of fins on a pipe of 12.7 mm diameter, for various materials (ref. [46]). Δt – temperature difference between pipe and environment, $t_r - t_f$; V_l – material volume per unit pipe length.

Example 8.5

When designing the equipment used in synthetic fibre production, the problem arises of minimizing the time during which the polymer is maintained in a liquid condition, in order to reduce the risk of thermal decay. The problem formulated below is concerned with the design of supply pipes connected to spinning nozzles.

Diameters are to be found for individual sections of the manifold supplying liquid polymer (Fig. 8.9) which give the shortest liquid residence time in the channels. The following data are known: liquid mass flow G, density ϱ, absolute viscosity μ, allowable pressure drop Δp; channel lengths l_1 and l_2.

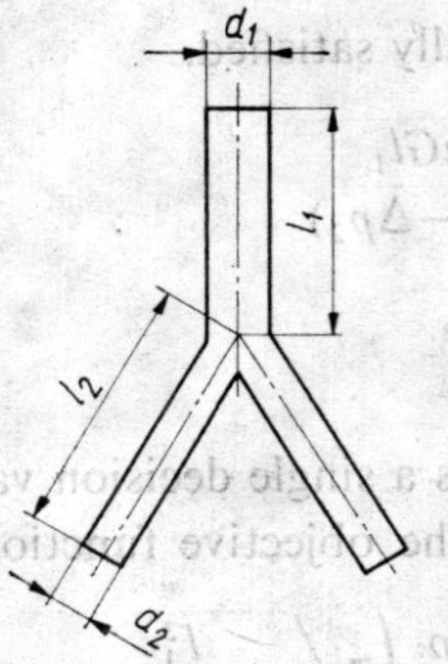

Fig. 8.9 Design sketch of a manifold.

Decision variables:
- channel diameters d_1 and d_2.

Parameters:
- channel lengths l_1 and l_2,
- liquid properties ρ and μ,
- mass flow G,
- pressure drop Δp.

The mathematical model should include a condition related to the allowable pressure drop (the indices denote manifold sections):

$$\Delta p_1 + \Delta p_2 = \Delta p \tag{8.33}$$

together with the equations expressing the relationships between pressure drops and channel dimensions. Owing to large liquid viscosity, the flow in the manifold is laminar. Neglecting pressure loss at channel branching (the loss is very small under laminar flow), Hagen–Poiseuille equation can be applied

$$\Delta p_1 = \frac{128\mu G}{\rho\pi d_1^4} l_1 \tag{8.34}$$

$$\Delta p_2 = \frac{128\mu 0.5G}{\rho\pi d_2^4} l_2 \tag{8.35}$$

The objective function is the total liquid residence time in the channels, calculated as the sum of quotients of channel volume and volume flow

$$f(d_1, d_2) = \frac{\pi d_1^2 l_1 \rho}{4G} + \frac{\pi d_2^2 l_2 \rho}{4 \times 0.5G}$$

The minimization of this function under constraint (8.33) is not very difficult, but the nonlinear constraint necessitates the use of a direct method. An alternative approach, based on a transformation of the variables using equations (8.33)–(8.35), ensures that in all subsequent substitutions constraint (8.33) is automatically satisfied.

$$d_1 = \sqrt[4]{\frac{128\mu G l_1}{\rho\pi(\Delta p - \Delta p_2)}} \tag{8.36}$$

$$d_2 = \sqrt[4]{\frac{64\mu G l_2}{\rho\pi\Delta p_2}} \tag{8.37}$$

Now, Δp_2 can be treated as a single decision variable and the inconvenient constraint is eliminated. The objective function can then be written as

$$f(\Delta p_2) = \frac{128\mu\pi\rho}{G}\left(\sqrt{\frac{l_1^3}{\Delta p - \Delta p_2}} + \sqrt{\frac{l_2^3}{0.5\Delta p_2}}\right)$$

The minimum can be determined by equating the function derivative with respect to Δp_2, to zero. The optimal value of Δp_2 is then

$$\Delta p_2^{\mathrm{opt}} = \frac{\sqrt[3]{2}\, l_2}{l_1 + \sqrt[3]{2}\, l_2}\Delta p$$

and the pressure drop in the first channel is

$$\Delta p_1^{\mathrm{opt}} = \Delta p - \Delta p_2^{\mathrm{opt}} = \frac{l_1}{l_1 + \sqrt[3]{2}\, l_2}\Delta p$$

Knowing the values of $\Delta p_1^{\mathrm{opt}}$ and $\Delta p_2^{\mathrm{opt}}$, the optimal channel diameters can be determined from equations (8.36) and (8.37).

In the above analysis, pressure drop has been introduced as an unknown variable in a purely formal way, by transformation of variables. More experienced readers, however, could probably at once identify the advantages of treating pressure drop as a decision variable and formulate the optimization problem accordingly.

8.3 SEALS

In process equipment design, the problem often appears of sealing the contact between adjoining stationary or moving elements. Such problems are of interest in that they require simultaneous consideration of relationships resulting from the mechanics of materials, heat and mass flow, friction and corrosion wear, etc., while also applying numerous design rules based on engineering practice. As malfunctioning seals can cause serious problems, time and effort devoted to producing a good design is easily justified. A common practice is to copy known solutions which have proven to be reliable (that is, hermetic), without paying much attention to their cost, energy consumption, etc. This practice is questionable as, for example, static seals may require more or less material, while dynamic seals may consume more or less energy.

Generally, the main features of optimization problems concerning seals are about the same as for the problems presented in Section 8.2 for heat transfer and fluid flow.

Example 8.6

The object here is to determine the width b and outer diameter D of the seal rings on a cylindrical piston. The rings are seated in grooves of diameter d (Fig. 8.10). Find the most advantageous number of rings, not less

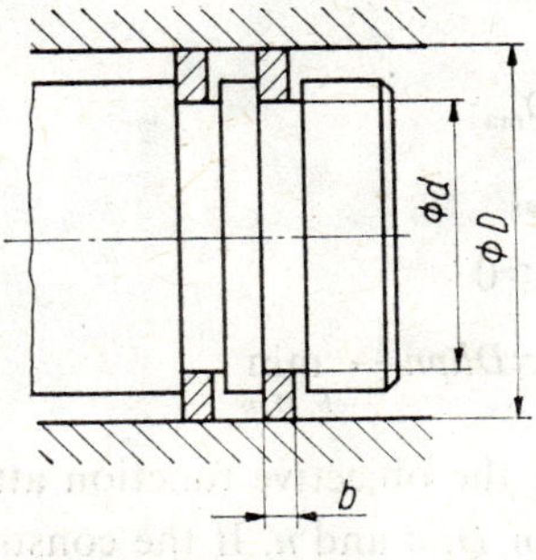

Fig. 8.10 Seal rings on a slider.

than two and not greater than four. Radial thrust on the seal rings is p, piston velocity is v. Owing to piston design limitations, the ring width must be contained between b_{min} and b_{max} and ring diameter between D_{min} and D_{max}.

Decision variables:

– ring width b, subject to

$$b_{min} \leqslant b \leqslant b_{max}$$

– ring diameter D, subject to

$$D_{min} \leqslant D \leqslant D_{max}$$

– number of rings n, subject to

$$2 \leqslant n \leqslant 4$$

Parameters:
– piston velocity v,
– radial thrust p,
– friction coefficient μ,
– allowable heat dissipation to the environment Q.

The mathematical model must include a constraint on heat dissipation at the seal

$$nPv \leqslant Q$$

where: $P = \mu\pi Dbp$ – friction force acting on one ring.

It is natural to consider, as an objective function, the total friction force acting on the pistcn. In a well-designed seal, this force should be as small as possible. The force depends on ring width and diameter and on number of rings

$$f(b, D, n) = nP = \mu\pi Dbpn$$

Optimal values of ring width and diameter can be found by solving the following problem

$$D_{\min} \leqslant D \leqslant D_{\max}$$

$$b_{\min} \leqslant b \leqslant b_{\max}$$

$$\mu\pi Dbpnv - Q \leqslant 0$$

$$f(b, D, n) = \mu\pi Dbpn \to \min_{b, D, n}$$

It can easily be seen that the objective function attains its minimum at the smallest possible values of D, b and n. If the constraint is satisfied, then the solution is $b = b_{\min}$, $D = D_{\min}$, $n = 2$.

CHAPTER 9

Shells

9.1 LEAST SURFACE AREA OF A BODY

The most economical shape of a shell for a piece of equipment is a problem frequently encountered in optimal design. Various problem forms have a common geometrical archetype: for a given capacity of a body of a known shape, the dimensions of the body are to be determined for which the outer surface area is the least.

In practical optimization problems, shell capacity need not be a design requirement. Instead, it is sometimes demanded that the shell should accommodate a certain structure which is necessary for the process. Also, one may come across a variety of evaluation criteria, such as shell weight or cost. Nevertheless, an appreciation of the nature of the geometrical problem makes it easier to formulate correctly its practical manifestation and to recognize the properties of the set of feasible solutions as well as of the objective function.

The example given below is a classical, purely geometrical problem of a surface area minimization. The decision variables represent body dimensions and a parameter represents body capacity; the mathematical model includes the relationship between capacity and dimensions, while the objective function represents the outer surface area of the body.

Example 9.1

Diameter D and height H are to be found for a cylinder of volume V, with least surface area.

Decision variables: D, H.

Parameter: V.

Constraint related to the volume

$$V - \frac{\pi}{4} D^2 H = 0$$

Objective function

$$f^*(D, H) = \pi DH + 2\frac{\pi}{4}D^2$$

and, after simplification

$$f(D, H) = DH + \frac{D^2}{2} \rightarrow \min_{D,H}$$

The simplest method of solution involves a substitution

$$H = \frac{4V}{\pi D^2}$$

which makes it possible to transform the objective function into the form

$$F(D) = \frac{4V}{\pi D} + \frac{D^2}{2}$$

This function attains its minimum at a point where the derivative with respect to D vanishes:

$$-\frac{4V}{\pi D^2} + D = 0$$

The optimal value of the diameter is thus

$$D_{\text{opt}} = \sqrt[3]{\frac{4V}{\pi}}$$

and the corresponding optimal height, from the formula used earlier in a substitution

$$H_{\text{opt}} = \sqrt[3]{\frac{4V}{\pi}} = D_{\text{opt}}$$

9.2 TANKS AND CONTAINERS

The essential idea of tank or container design is to form a shell of a specific shape so as to make it hold a given volume or to exhibit some other properties. The optimization problem may be concerned with designing a shell of minimum weight or of minimum cost. The difference between this and the problem of minimum surface area is that an account must be taken of the relations known from the theory of strength of materials concerning shell thickness, allowable strain, etc. In the case of atmospheric pressure tanks, designing a shell of minimum weight or minimum cost can sometimes be reduced to the purely geometrical problem of finding the least shell surface area.

On the other hand, pressure tank design generally involves process-related relations; therefore, process quantities – especially pressure and temperature – may appear as parameters in problem formulation.

The idea of the most advantageous shape of pressure vessels has for long been known [9, 19, 85]. It is still common, however, to fail to consider all possible questions because practical design requirements may differ widely, as illustrated in the examples below.

Example 9.2

The dimensions of a thin-walled, slender stainless steel container closed with a piston are to be found. Shell weight per unit volume of the container should be minimal. Wall thickness must not be less than $g_{min} = 0.25$ mm. In order to keep the closure leakproof, radial deflection of the shell cannot exceed $\delta_{max} = 0.025$ mm. Pressure inside the container is $p = 16.8$ MPa, allowable stress $k = 400$ MN/m², Young's modulus $E = 2 \times 10^5$ MN/m², Poisson's ratio $\nu = 0.33$.

Decision variables:

– wall thickness g, subject to $g \geqslant g_{min}$,

– diameter d.

Parameters:

– material constants k, E, ν;

– pressure p;

– allowable radial deflection δ_{max}.

The mathematical model consists of inequalities reflecting the requirements of shell proportions qualifying it as thin-walled, allowable radial deflection and allowable combined stress:

$$\frac{d}{g} \geqslant 20, \qquad \frac{d}{2E}\left(\frac{pd}{2g} - \nu \frac{pd}{4g}\right) \leqslant \delta_{max}$$

$$\sqrt{\left(\frac{pd}{4g}\right)^2 - \frac{pd}{4g}\,\frac{pd}{2g} + \left(\frac{pd}{2g}\right)^2} \leqslant k$$

The objective function, i.e. the ratio of shell weight to container volume can be expressed by a formula neglecting the weight of the container bottom as its share is negligible in the weight of a very slender cylinder

$$f^*(d, g) = \frac{\rho \pi d L g}{\dfrac{\pi d^2}{4} L}$$

where: ρ – material density.

Having transformed constraints and objective function and substituted parameter values, we obtain the following problem form

$$20g - d \leqslant 0$$

$$d - 7.422\sqrt{g} \leqslant 0$$

$$\frac{d}{g} - 55 \leqslant 0$$

$$f(d, g) = \frac{g}{d} \rightarrow \min_{d,g}$$

The graphical representation of this problem is shown in Fig. 9.1. Its peculiarity lies in the existence of multiple solutions described by the equation

$$\frac{d}{g} = 55$$

where g can assume values between 0.25 and 0.44 mm (segment AB). This is a consequence of the fact that the boundary of the feasible region, as defined by the allowable stress constraint, coincides with the objective function contour.

It should be pointed out that the qualitative properties of the solution may alter as parameter values are varied. If, for example, another steel grade characterized by larger allowable stress value is taken, then the slope

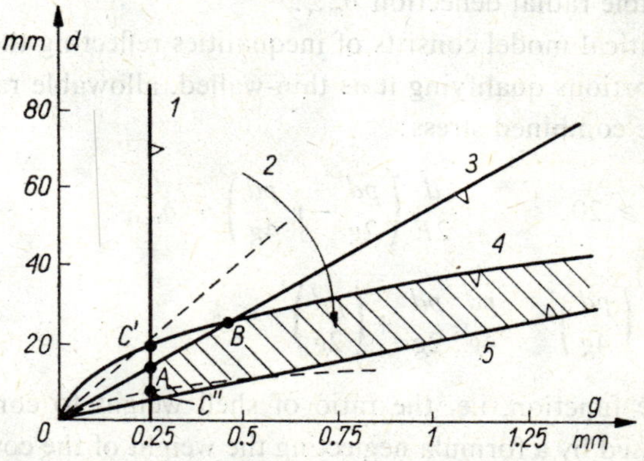

Fig. 9.1 Graphical representation of the problem of container shell optimization. Segment AB is a locus of optimal solutions. Dashed lines and points C′ and C″ represent the positions of constraints and optimal solutions corresponding to modified parameter values (see discussion in the text). 1 – least thickness constraint, 2 – the direction of increasing values of the objective function, 3 – allowable stress constraint, 4 – allowable deflection constraint, 5 – shell proportions constraint.

of the line representing the stress constraint (dashed line) increases. At a slope greater than a certain critical value, this constraint clearly becomes redundant and the minimum value of the objective function is found at the crossing point of the other two constraints (point C′). Similarly, if the allowable radial deflection δ_{max} is reduced, the curve representing deflection constraint is shifted to a lower position (dashed curve, minimum at point C″). It can be easily seen that reducing δ_{max} still further may cause the problem to become insoluble: if the curve is shifted to a position below the line representing the shell proportions constraint, then the set of feasible solutions is empty.

Example 9.3 [39]

The problem is to find the most advantageous – with respect to weight – series of main dimensions of condensate tanks of 1–12.5 m³ volume, to be used in the heating systems of sugar plants.

A condensate tank, also known as a hot-well, is a cylindrical pressure vessel with ellipsoidal heads, typically made of steel sheet of uniform thickness, equipped with nozzles, manhole, water-level gauge, supports, etc. (Fig. 9.2). Rated pressure is 0.4 MPa.

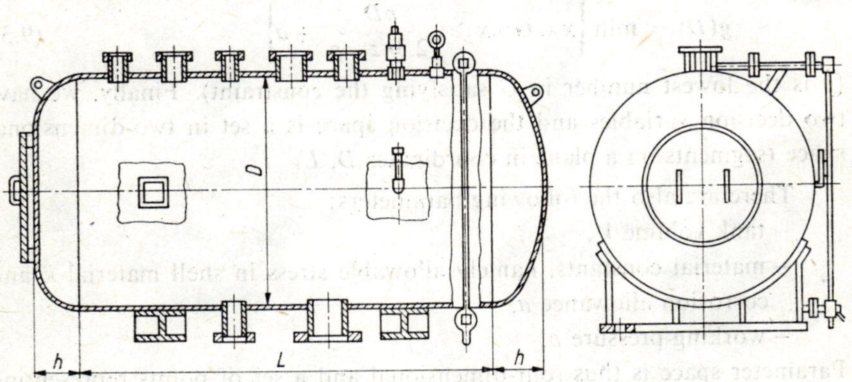

Fig. 9.2 A condensate tank for the heating system in a sugar plant.

Tank weight analysis can be based on three decision variables: length of cylindrical part L, diameter D and shell thickness g. According to Polish National Standards defining the dimensions of ellipsoidal heads, if we take into account an additional requirement that the manhole in the tank head should be large enough, then diameter D has to be chosen from the series {1000, 1200, 1400, …, 2200 mm}. As the tanks are installed in buildings with a typical planning module of 6 m, total tank length must not exceed 5 m.

According to the Polish pressure vessel code, tank wall thickness in the cylindrical section must satisfy the inequality

$$g \geqslant \frac{pD}{2.3kz-p}+a \tag{9.1}$$

where: z – reinforcement coefficient accounting for openings and nozzles, a – corrosion allowance.

Head thickness is subject to the condition

$$g \geqslant \frac{pDy}{4k}+a \tag{9.2}$$

where: y – characteristic head coefficient. For typical condensate tanks, we have from the vessel code: $z = 0.7$ and $y = 2$. Constraint (9.1) is thus more rigorous than (9.2) and the latter can be disregarded. Steel sheet thickness must be chosen from a set of values defined by the Polish National Standards:

$$g \in G, \qquad G = \{4, 5, 6, 7, 8, \ldots, \text{mm}\}$$

It follows from (9.1) that instead of treating thickness g as a decision variable, we can consider it as a function of diameter D, function values belonging to set G

$$g(D) = \min\left\{x \in G;\ x \geqslant \frac{pD}{2.3kz-p}+a\right\} \tag{9.3}$$

(g is the lowest number in G satisfying the constraint). Finally, we have two decision variables and the decision space is a set in two-dimensional space (segments on a plane in coordinates D, L).

There are also the following parameters:

– tank volume V,
– material constants, namely allowable stress in shell material k and corrosion allowance a,
– working pressure p.

Parameter space is thus four-dimensional and a set of points representing tank volumes is given in that space. For every point, an optimal decision should be found. In the mathematical model, an approximate relation appears between tank volume and its main dimensions:

$$V = \pi\left(\frac{D^2L}{4}+\frac{h(D)}{24}\{3D^2+4[h(D)]^2\}\right) \tag{9.4}$$

where: $h(D)$ – head depth defined by the relevant Polish National Standard (also given in Table 9.1). With an accuracy of $\pm 2\%$, it can be approximated by the formula

$$h(D) = 0.07825\exp(1.554D-0.2954D^2) \tag{9.5}$$

Table 9.1
Height of ellipsoidal tank heads as a function of tank diameter

D, mm	700	800	900	1000	1100	1200	1300
h, mm	200	225	250	275	300	325	375
D, mm	1400	1500	1600	1700	1800	1900	2000
h, mm	390	415	440	465	490	515	540

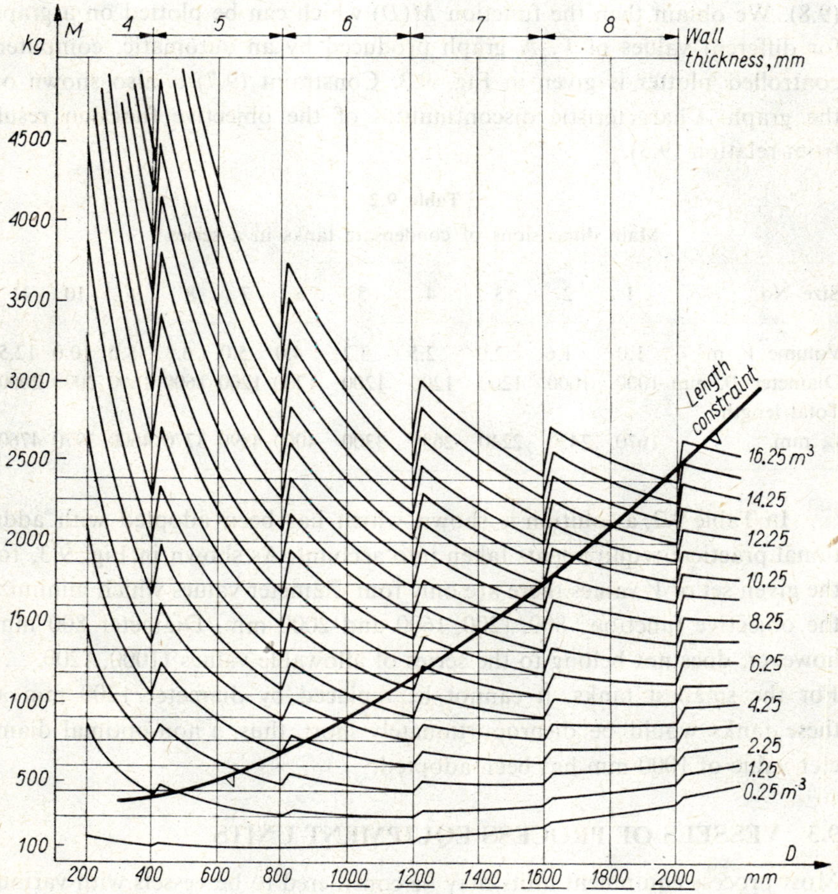

Fig. 9.3 Tank mass as a function of diameter and volume. The feasible region lies below the curve representing length constraint.

Upon transformation of (9.4), the length of the cylindrical section can be expressed as:

$$L(D) = \frac{4}{D^2}\left(\frac{V}{\pi} - \frac{h(D)}{24}\{3D^2+4[h(D)]^2\}\right) \tag{9.6}$$

The constraint on total tank length can be written as

$$L(D)+2h(D) \leqslant 5 \tag{9.7}$$

The objective function, i.e. the weight of tank shell

$$M(D,L) = \rho g(D)\pi\{DL+D^2+2[h(D)]^2\} \tag{9.8}$$

where: $g(D)$ – function defined by formula (9.3).

The simplest method of solution is to substitute formula (9.6) into (9.8). We obtain then the function $M(D)$ which can be plotted on a graph for different values of V. A graph produced by an automatic, computer-controlled plotter is given in Fig. 9.3. Constraint (9.7) is also shown on the graph. Characteristic discontinuities of the objective function result from relation (9.3).

Table 9.2

Main dimensions of condensate tanks in a series

Size No.	1	2	3	4	5	6	7	8	9	10	11
Volume V, m³	1.0	1.6	2.0	2.5	3.2	4.0	5.0	6.3	8.0	10.0	12.5
Diameter D, mm	1000	1000	1200	1200	1200	1200	1200	1600	1600	2000	2000
Total length L, mm	1670	2430	2240	2680	3300	4000	4890	3770	4600	3970	4760

In Table 9.2, a solution is shown which has been adopted with additional practical requirements taken into account. As shown in Fig. 9.3, for the given set of V values there are only four diameter values which minimize the objective function: 800, 1200, 1600 and 2000 mm. Diameter 800 mm, however, does not belong to the series of allowable values (1000, 1200, ...). For the smallest tanks, it cannot be replaced by diameter 1200 mm as these tanks would be disproportionately short, thus, a non-optimal diameter value of 1000 mm has been adopted.

9.3 VESSELS OF PROCESS EQUIPMENT UNITS

Most process equipment units may be considered to be vessels with various necessary modifications enabling the units to perform the required functions. Vessel optimization with respect to weight or cost is essentially a generalization of the container optimization problem. Characteristically, process quantities appear as parameters or even as decision variables in the problem and the mathematical model involves relations between process parameters as well as descriptions of the phenomena occurring in the process (see Examples 9.4 and 9.5). In some cases, however, the process description is unnecessary. This applies to, among others, the most economical proportioning of vessels in process equipment units of such design as can be treated as practically validated and operable, i.e. when the reliable technological and economic data, needed for designing the units, exist (Example 9.6).

Example 9.4 [79, 81, 82]

The problem is to find the dimensions which minimize the weight of a pressurized Robert-type evaporator (Fig. 9.4), in which an aqueous solution of known properties is concentrated. The evaporator is heated by saturated steam.

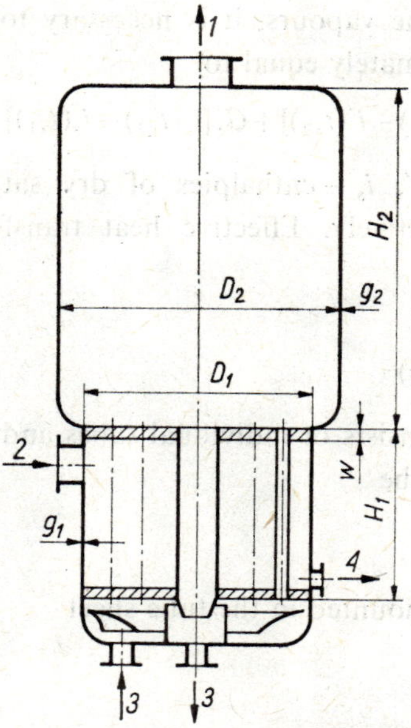

Fig. 9.4 Schematic longitudinal cross-section of a Robert-type evaporator. 1 – vapour, 2 – heating steam, 3 – solution, 4 – condensate.

Before an optimization problem is formulated, let us summarize the most important facts about the equipment unit considered. Design quantities characterizing a Robert-type evaporator are: diameter and height of the heating chamber shell, D_1 and H_1, diameter and height of the vapour chamber shell, D_2 and H_2, thicknesses of each shell g_1 and g_2, diameter of heating tubes d, pitch in a hexagonally arranged tube bundle u, tube sheet thickness w, tube sheet area utilization ratio r, allowable vapour velocity in vapour chamber c. As the evaporator shell is thin-walled, no distinction is made hereafter between inner and outer diameters of both shell parts. It is also assumed that d, u, w and r are not subject to designer's decision.

The major process quantities are: solution concentration at inlet b_1 and outlet b_2, solution temperature at inlet t_{s1} and outlet t_{s2}, heating steam temperature t, mass flow of solution at inlet G_s, heat transfer coefficient k, heat loss ratio q.

Mass flow of vapour, G, generated while concentrating the solution is given by

$$G = G_s\left(1 - \frac{b_1}{b_2}\right)$$

In order to generate the vapours, it is necessary to deliver, in unit time, a heat amount approximately equal to

$$Q = \{G[i''(t_{s2}) - i'(t_{s2})] + G_s[i_s(t_{s2}) - i_s(t_{s1})]\}(1+q)$$

where: functions i'', i', i_s – enthalpies of dry saturated steam, boiling water, solution, respectively. Effective heat transfer requires a heating surface area

$$F = \frac{Q}{k(t - t_{s2})} \tag{9.9}$$

The heating surface consists of individual tubes and the tube sheet surface area allotted to each tube

$$s = 0.866u^2$$

The number of tubes mounted in the tube sheet

$$n = \frac{\pi D_1^2 r}{4s} \tag{9.10}$$

The tubes make the required heating surface area

$$F = n\pi d H_1 \tag{9.11}$$

For stability reasons, the tubes must be shorter than certain allowable length $H_{1\max}$

$$H_1 \leqslant H_{1\max}$$

As vapour velocity in vapour chamber must be smaller than its allowable value c, a condition is obtained that

$$D_2 \geqslant \sqrt{\frac{4Gv''(t_{s2})}{\pi c}} \tag{9.12}$$

where: v'' – specific volume of dry saturated steam.
An obvious design condition is

$$D_2 \geqslant D_1$$

Introducing the notation S for the vapour chamber cross-section area,

$$S = \frac{\pi D_2^2}{4}$$

a constraint on vapour chamber height, resulting from the requirement of efficient separation of solution droplets, can be formulated

$$H_2 \geqslant \frac{G}{S\phi(p)0.276} \tag{9.13}$$

where: ϕ – empirical function of vapour pressure p, which is equal to saturation pressure at temperature t_{s2}

$$p = p_s(t_{s2})$$

The thickness of the heating chamber shell should be suited to a working pressure which is a function of temperature (saturation pressure at temperature t):

$$P = p_s(t)$$

and thus, according to the Polish pressure vessel code

$$g_1 \geqslant \frac{PD_1}{2.3k - P} + a \tag{9.14}$$

where: k – allowable stress in shell material, a – corrosion allowance.

For the sake of design simplification, it is further assumed that $g_2 = g_1$, even though in pressure evaporators (contrary to vacuum ones) the load on vapour chamber walls is smaller than on heating chamber walls.

In can be concluded from the relations specified above that decision variables are g_1, D_1, D_2, H_1, H_2 and parameters are: d, u, w, r, c, b_1, b_2, t, t_{s1}, t_{s2}, G_s, q, k. As wall thickness must be chosen from a set of values as defined by a standard on sheet steel, decision space is a set of hyperplanes in a five-dimensional Euclidean space. Parameter space in which design data define a single point is a twelve-dimensional Euclidean space.

As suggested by the initial problem statement, the objective function has the meaning of evaporator weight. It should be pointed out, however, that the weight of tubes in the heating surface is independent of major evaporator dimensions because heating surface area is defined by the parameters. (The weight of heating tubes of known diameter making a definite heating surface area is always the same, irrespective of tube length). Major dimensions affect, though, the weight of the heating chamber shell. Treating the heads as circular plates, the variable part of the evaporator weight can be approximately expressed as

$$M(g_1, D_1, D_2, H_1, H_2) = \rho\left\{\pi g_1\left[\left(\frac{D_1^2}{4}+\frac{D_2^2}{4}\right)+D_1H_1+D_2H_2\right]+ +2nw\left(s-\frac{\pi d^2}{4}\right)\frac{1}{r}\right\}$$

where: ρ – shell material density.

The formula above can be adopted as a definition of the objective function. The relations given earlier make it possible to optimize evaporator designs at different values of the parameters. In practice, as vapour velocity in the vapour chamber is usually small, it is often assumed that $D_1 = D_2$. Height H_2 depends then on diameter D_1 and finally, as thickness g_1 has been defined as a function of D_1, only H_1 and D_1 remain as decision variables. In Fig. 9.5, a relation between the variable part of evaporator

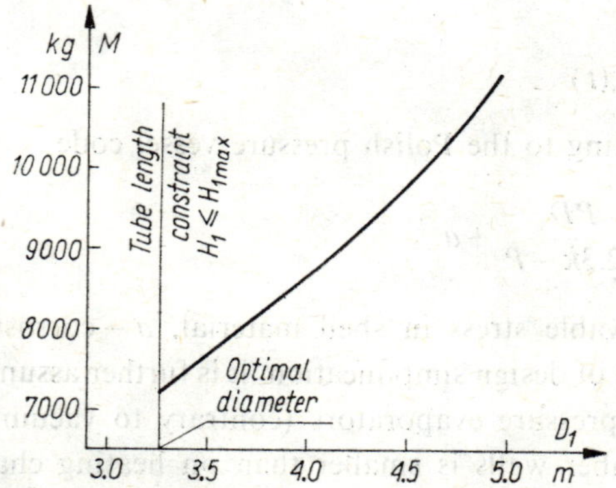

Fig. 9.5 Variable part of evaporator mass as a function of heating chamber diameter D_1 (constant heating surface area 1500 m^2; this is a real evaporator design destined for sugar plants).

weight and diameter D_1 is shown together with optimal diameter, for arbitrarily chosen parameter values which are equivalent to a heating surface

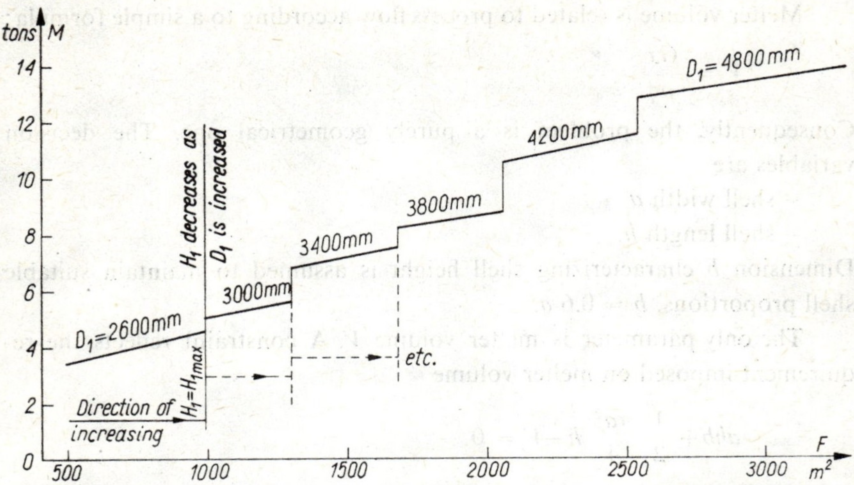

Fig. 9.6 Variable part of evaporator mass at optimal heating chamber diameter (standardized diameter values only), as a function of heating surface area.

area of 1500 m^2. Fig. 9.6 shows the result of the optimization of diameter D_1 for several evaporator units of various heating surface areas (the set of D_1 values is defined by the Polish Standard on ellipsoidal heads). As explained in the graph, the discontinuity points in the curve of smallest possible weight are due to the necessity to avoid a tube length greater than the allowable value $H_{1\max}$.

Example 9.5 [80]

Weight-optimal dimensions are to be found for the shell of a melter, i.e. a unit in which sugar is melted in water or in beet juice (Fig. 9.7). The melter capacity, i.e. mass flow of sugar solution required is G. The data

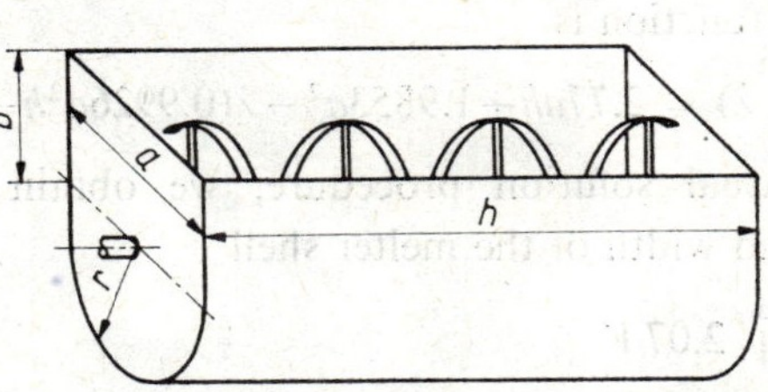

Fig. 9.7 Design sketch of a melter. The shaft is equipped with a ribbon stirrer.

given is the time required for sugar to dissolve τ, and solution density ρ. Possible fluctuations in the solution mass flow should be accounted for by using a load ratio q.

Melter volume is related to process flow according to a simple formula:

$$V = \frac{G\tau}{\rho q}$$

Consequently, the problem is a purely geometrical one. The decision variables are

– shell width a

– shell length h.

Dimension b characterizing shell height is assumed to maintain suitable shell proportions, $b = 0.6\, a$.

The only parameter is melter volume V. A constraint reflects the requirement imposed on melter volume

$$ahb + \frac{1}{2}\frac{\pi a^2}{4}h - V = 0$$

Having substituted $b = 0.6\, a$, we obtain

$$0.9926\, a^2 h - V = 0$$

Shell thickness is determined on the basis of design experience, taking shell stability and corrosion allowance into account. Consequently, shell surface area – instead of shell weight – can be accepted as the objective function:

$$f(a, h) = \left(2b + \frac{\pi a}{2}\right)h + 2\left(ab + \frac{\pi a^2}{8}\right)$$

and, after eliminating b,

$$f(a, h) = 2.77ah + 1.9853a^2$$

The problem can be solved by the Lagrange multiplier method. The augmented objective function is

$$f_a(a, b, \lambda) = 2.77ah + 1.9853a^2 - \lambda(0.9926a^2 h - V)$$

Following the usual solution procedure, we obtain formulae defining optimal length and width of the melter shell

$$h_{\text{opt}} = \sqrt[3]{2.07\, V}$$

$$a_{\text{opt}} = \sqrt[3]{0.70\, V}$$

Example 9.6

The problem is to find optimal dimensions of a floating-head type heat exchanger, Fig. 9.8. The objective function is the manufacturing cost –

which is to be minimized for a given transfer surface area, F. A series of exchangers of the same design has been manufactured over a period of several years, so detailed cost data are available for all structural parts.

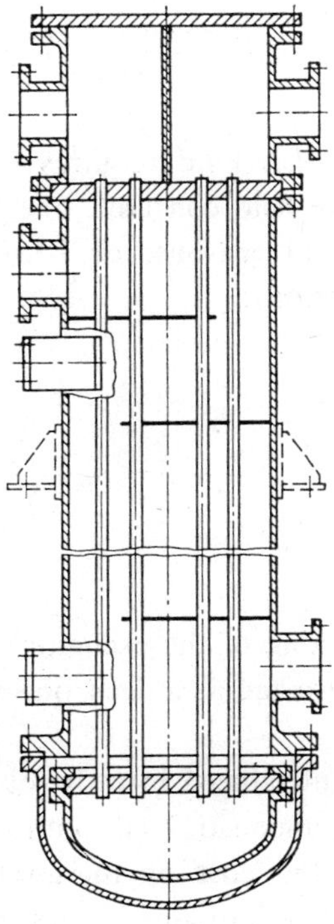

Fig. 9.8 Longitudinal cross-section of a floating-head type tubular heat exchanger.

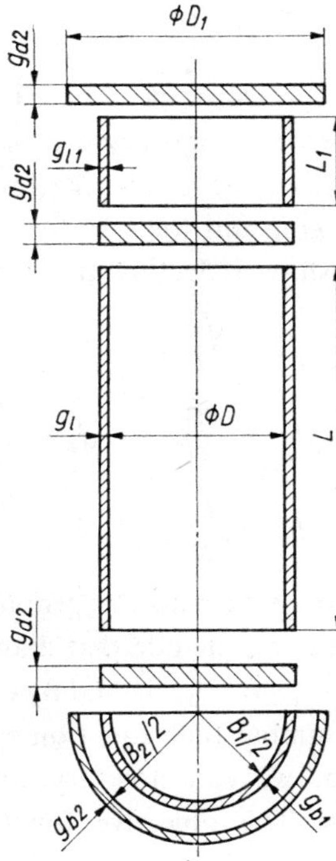

Fig. 9.9 Parts of exchanger structure (geometrically simplified).

The exchanger structure can be approximately treated as a set of parts shown in Fig. 9.9. As expected, the dimensions of the optimal structure do not differ very much from the dimensions of the structures manufactured earlier. Accordingly, unit costs (per 1 kg weight) of the parts in the optimal structure will remain unchanged.

Prior to the formulation of the optimization problem, it is advisable to consider some fundamental relations. Let d denote the diameter of the heating tubes, s tube sheet surface area per tube and r tube sheet surface area utilization ratio. The heat transfer surface area is

$$F = \frac{\pi D^2}{4rs} \pi d L$$

Hence, feasible heat exchanger designs must satisfy the equation

$$L = F\frac{4rs}{\pi^2 d}\frac{1}{D^2} \tag{9.15}$$

As F, r, s and d are known, the equation can be rewritten as

$$L = \frac{C}{D^2} \tag{9.16}$$

where: C – constant.

Correct proportioning of a structure of this kind requires that for a given exchanger size, the following ratios are held constant, or approximately constant (either for purely geometrical reasons, or because of relationships reflecting the principle of operation):

$$l_1 = \frac{L_1}{L},$$

$$b_1 = \frac{B_1}{D}, \quad d_1 = \frac{D_1}{D},$$

$$b_2 = \frac{B_2}{D}, \quad d_2 = \frac{D_2}{D} \tag{9.17}$$

If we now return to the problem of proportioning of the exchanger structure, we can conclude that diameter D can be treated as a decision variable while C, l_1, d_1, d_2, b_1 and b_2 are parameters.

Assuming that equation (9.16) could be introduced into the objective function, we can now try to formulate an unconstrained optimization problem. The objective function expressing the cost of the exchanger design must rely on knowledge of the unit cost of individual parts, respectively: k_l, k_{l1}, k_{d1}, k_{d2}, k_{b1}, k_{b2}. Then, the thicknesses of individual parts can be identified as functions of their characteristic dimensions, $g_l(D)$, $g_{l1}(D)$, $g_{d1}(D)$, $g_{d2}(D)$, $g_{b1}(B_1)$, $g_{b2}(B_2)$. Proper formulae are either defined in pressure vessel code, or are known from general mechanics of materials; for example, the first function is a relation of the form (9.3). Finally, the cost can be expressed by a formula representing the sum of the costs of individual parts:

$$f(D) = \rho\pi\left[k_l DLg_l(D) + k_{l1} DL_1 g_{l1}(D) + k_{d1}\frac{D_1^2}{4}g_{d1}(D) + \right.$$
$$\left. + k_{d2}\frac{D_2^2}{2}g_{d2}(D) + k_{b1}\frac{B_1^2}{2}g_{b1}(B_1) + k_{b2}\frac{B_2^2}{2}g_{b2}(B_2)\right]$$

Note that the total cost depends on diameter D only, as length L can be eliminated using formula (9.16) and the remaining dimensions can be eliminated using formulae (9.17).

The function $f(D)$ can most conveniently be minimized graphically, as this yields the optimal solution while also making it possible to carry out a sensitivity analysis. Another important factor is that the functions expressing part thicknesses are generally discontinuous because (as in Example 9.3) function values must belong to the set of standardized sheet steel thickness values.

The above analysis is based on the methods described in [23] which have been applied in the Soviet process equipment industry. The prerequisite for the success of a similar optimization approach to any heat exchanger design is the availability of adequate cost data, from which cost relations can be identified for individual parts. This is not always the case with practical cost data processing and storage procedures.

CHAPTER 10

Heat transfer equipment

10.1 HEAT EXCHANGERS

There are at least ten types of heat exchangers of different designs that can be considered for process engineering applications. The choice of a specific exchanger is an important decision in itself and may also be optimized. Usually, such a decision is made on the basis of practical selection rules or is aided by a comparison of the values of some approximate indices [36, 76]. Once a particular type of heat exchanger has been chosen, then an optimization problem can be formulated in a similar manner to that shown for other process equipment units.

Heat exchangers find application most often in complex systems, where thermal processes are coupled with other processes. If we consider the task a heat exchanger performs concerning the energy flow in the system, then the design of the exchanger may require consideration of system decomposition; if not by accurate methods, then at least according to the rules discussed in Section 3.3. It is thus necessary to establish the number n of process quantities defining connections between the exchanger and the system, and the number m of equations in the mathematical model of the process; the model may include: the equations of energy conservation (overall energy balance in the exchanger and partial balances concerning its parts), the equation expressing the kinetics of heat transfer, hydraulic pressure loss, etc. As pointed out in Section 3.3 for any type equipment unit, for heat exchanger design also, different formulations of the requirements as to the process effects are possible. Depending upon the number s of given process quantities, the optimization of a heat exchanger may assume one of two basically different forms which, henceforward, will be called problems P1 and P2 (as proposed in [47]). Problem P1 corresponds to a case where $s = n-m$, while in P2 $s < n-m$. In problem P1, the exchanger can be virtually separated from the system, as all process quantities are uniquely defined. On the other hand problem P2 involves some process quantities (their number is $n-m-s$) which should be considered as decision variables.

Characteristic features of problem P1 are as follows:

– the decision variables correspond to geometrical properties of the exchanger design (dimensions, the number of multiple elements, etc.);

– parameters correspond to process quantities, physical properties of heat-transfer media, structural materials and characteristic figures of design requirements;

– mathematical models include geometrical relations, those of heat transfer and fluid flow as well as formulae involving mechanics of materials;

– objective functions have the meaning of various design indices; this could be one of the structural, process-dependent or process/economic criteria (as discussed in Chapter 2).

As an example of a problem of type P1, the minimum weight design of a heat exchanger (Example 3.3) may be cited.

It should be added that there is a lack of consensus on suitable optimization criteria for heat exchanger design. An opinion prevails that structural and thermodynamic indices should not be generally used in a design evaluation as they cannot reflect such practically important factors as the utilization of rated capacity or energy consumption for the pumping of fluids. Such indices may therefore be used as auxiliary criteria only. The basic design evaluation measure would normally be some economic index. It should be borne in mind, however, that there are problems in which no reliable economic data can be identified and thus thermodynamic indices must be used as the design evaluation tools.

Heat exchanger structures are often assembled from standardized parts, like tubes, tube sheets, heads, nozzles, etc. As a result, discrete variables appear in decision-making problems and it is generally recognized that the presence of such variables induces numerical difficulties. If we consider a series of designs as defined by standards or catalogues, then some discrete variables depend on others; it is thus possible to separate them into two groups referred to as *leading variables* and *dependent variables* [47]. It is sufficient to include, among the decision variables, the leading discrete variables only, thus reducing the problem dimensionality (Fig. 10.1). This technique can conveniently be used, even if the relation between dependent variables and leading variables is not unambiguous, that is if several values of the dependent variables can be assigned to one vector of leading variables.

Problems of type P2 involve simultaneous optimization of the exchanger structure and, to some extent, of the process. A good example of a problem of this kind is the optimization of a cooler operated in an open cooling circuit attached to a process plant (Fig. 10.2). For correct plant operation,

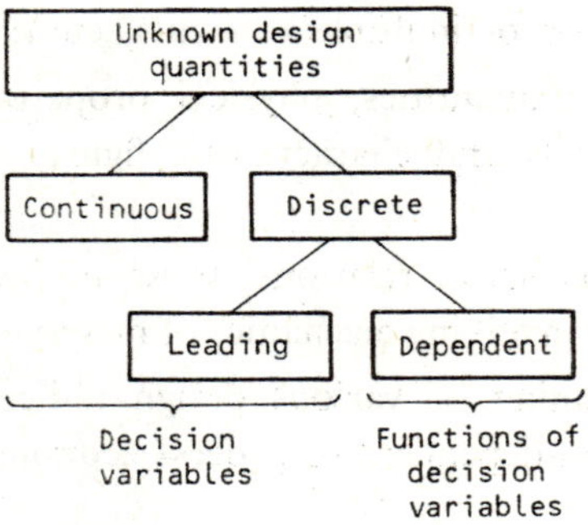

Fig. 10.1 Partition of design quantities appearing in optimization problems featuring standardized equipment parts.

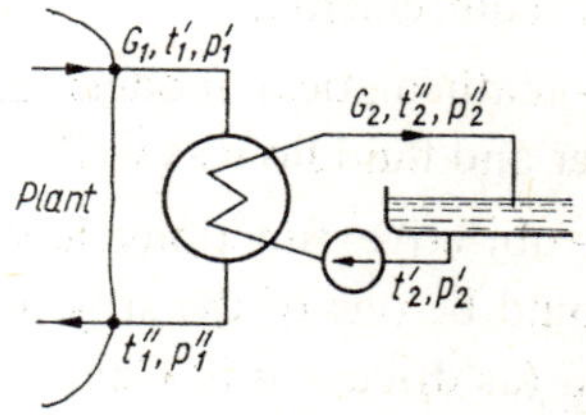

Fig. 10.2 A cooler operated in an open loop.

it is necessary to maintain certain temperatures and pressures in the cooled fluid t_1', t_1'', p_1', p_1'' at the given mass flow rate G_1 and these are the parameters in the optimization problem. Other parameters are: initial temperature and pressure of the cooling agent t_2', p_2'. Optimization is concerned with cooler structure, mass flow rate G_2, outlet temperature t_2'' and outlet pressure p_2''.

Problem P2 can be usually stated in a mathematical form characteristic of nonlinear programming problems. Its formal properties are similar to the properties of P1 except for the presence of process quantities among the decision variables. The only type of objective function which is really suited for problems P2 is, therefore, a function representing total cost (capital and operation) of the plant part considered, or representing a certain economic index based on the components of the total cost.

Example 10.1

The problem is to find the optimal design of a heat exchanger assembled from standardized 'double-pipe' segments (Fig. 10.3). Standard data on such segments are summarized in Table 10.1. The heat exchanger is to cool down toluene flowing at a mass flow rate $G = 5.3$ kg/s from $t_1' = 100°C$ to $t_1'' = 50°C$. Water is employed as the cooling agent at a mass flow rate of $G_2 = 4$ kg/s and an initial temperature $t_2' = 15°C$. The pressure drop in each liquid must not exceed $\Delta p_{1\max} = 0.05$ MPa and $\Delta p_{2\max} = 0.08$ MPa,

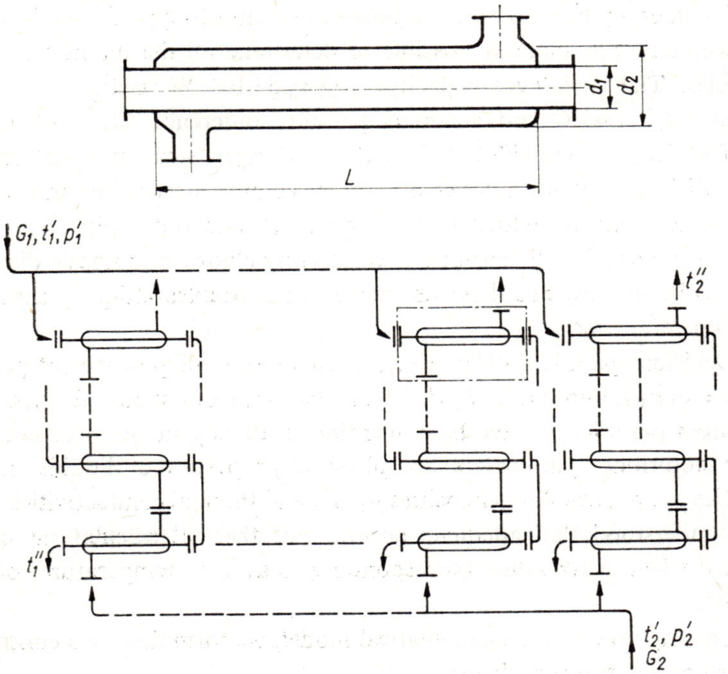

Fig. 10.3 Scheme of a double-pipe heat exchanger; above, design principle of a single segment.

Table 10.1

Standardized segments of double-pipe heat exchangers

Inner pipe Diameter d_1 mm	Inner pipe Thickness g_1 mm	Outer pipe Diameter d_2 mm	Outer pipe Thickness g_2 mm	Length L m
38	2.5	57	3	3, 4, 6
57	3	89	3.5	3, 4, 6
89	3.5	133	4	4, 6
108	4	159	4.5	6

respectively. The optimal design should minimize total annual cost calculated under the following assumptions: annual utilization time $\theta = 7000$ h, power cost $\varkappa_e = 0.002$ £/kWh, pumping efficiency $\eta = 0.75$, unit cost of material (carbon steel) $\varkappa_m = 0.2$ £/kg, life time $\tau = 6$ years. Average thermal resistivity of fouling deposits at heat transfer surface: toluene $r_1 = 0.00018$ m^2 K/W, water $r_2 = 0.00035$ m^2 K/W. Thermal conductivity of steel $\lambda_s = 52$ W/(mK).

It can easily be seen that this problem belongs to type P1, as the energy balance equations make it possible to determine all the unknown process variables. The exchanger capacity is $Q = 5 \cdot 10^5$ W, outlet temperature of water $t_2'' = 45°C$ and mean temperature difference $\Delta t_m = 44$ K.

The decision variables definig the exchanger structure are: number of parallel sections m, number of serially connected elements in a section n, element length L, diameter of inner pipe d_1 and outer pipe d_2. Thicknesses of pipe walls, dimensions of connecting elbows and nozzle diameters are also important, but these are dependent variables uniquely associated with pipe diameters.

Decision space is a discrete set, as m must be a positive integer, n a positive even number and L, d_1 and d_2 chosen from standard sizes.

Most parameters have been mentioned already in the problem statement: additional parameters are physical properties of liquids, namely densities ρ_1, ρ_2, absolute viscosities μ_1, μ_2 and thermal conductivities λ_1, λ_2 (it is understood that adequate accuracy of thermal calculations can be attained when using values corresponding to average temperatures of both liquids).

Coming now to the mathematical model, we formulate two constraints concerning the pressure drop:

$$\Delta p_1(x) \leqslant \Delta p_{1\max} \tag{10.1}$$

$$\Delta p_2(x) \leqslant \Delta p_{2\max} \tag{10.2}$$

where: functions Δp_1 and Δp_2 are given in a nomograph (cf. Appendix 4). Much more complicated is the formulation of a constraint concerning the thermal capacity of the exchanger

$$Q \leqslant k(x)\Delta t_m F(x) \tag{10.3}$$

where: F – heat transfer surface area

$$F(x) = \pi d_1 L n m$$

k – heat transfer coefficient [70].

$$k(x) = \frac{1}{\beta_1 \cdot \dfrac{1}{\alpha_1} + \dfrac{d_1}{2\lambda_r} \ln \beta_1 + r + \dfrac{1}{\alpha_2}}$$

and λ_r – thermal conductivity of the material of inner pipe wall. Other quantities are defined by the following formulas:

$$\beta_1 = \frac{d_1}{d_1 - 2g_1}$$

$$r = r_1 \frac{d_1}{d_1 - 2g_1} + r_2$$

$$\alpha_1 = 0.021 Re_1^{0.83} Pr_1^{0.43} \frac{\lambda_1}{d_1 - 2g_1} \tag{10.4}$$

$$\alpha_2 = 0.021 Re_2^{0.83} Pr_2^{0.43} \frac{\lambda_2}{d_2 - 2g_2 - d_1} \tag{10.5}$$

where: r_i – thermal resistivity of fouling deposits on the heat transfer surface; Re_i, Pr_i – Reynolds and Prandtl numbers characterizing flow in the exchanger channels, $i = 1$ denoting inner and $i = 2$ the outer channel. Reynolds numbers in both channels depend on decision variables:

$$Re_1 = \frac{4G_1}{\pi m \mu_1 (d_1 - 2g_1)}$$

$$Re_2 = \frac{4G_2}{\pi m \mu_2 (d_2 - 2g_2 + d_1)}$$

(Formulae (10.1) and (10.2) are true if $Re_1 > 800$ and $Re_2 > 50\,000$, respectively, otherwise, alternative formulae should be used [70]).

The objective function can be summarized as

$$f(\boldsymbol{x}) = \frac{M(\boldsymbol{x})\varkappa_m}{\tau} + C(\boldsymbol{x}) \tag{10.6}$$

where: M – exchanger weight, C – cost of power consumed in pumping of fluids

$$C(\boldsymbol{x}) = \left[\frac{\Delta p_1(\boldsymbol{x}) G_1}{\rho_1} + \frac{\Delta p_2(\boldsymbol{x}) G_2}{\rho_2} \right] \frac{\theta \varkappa_e}{\eta} \tag{10.7}$$

The problem formulation is now complete: formulae (10.1), (10.2) and (10.3) represent constraints while formulae (10.6) and (10.7) make it possible to calculate the objective function. Such a problem can be solved by the general numerical methods of nonlinear programming [47, 70], or by a simple method suited to non-automated calculations. This method is based on the observation that the objective function is unimodal (cf. Section 4.4) with respect to variables m, d_1 and d_2. The explanation of this property involves the fact that at given values of mass flows of both fluids, m, d_1 and d_2 unambigously define flow velocities. On the other hand, it can be easily seen that the objective function is unimodal with respect to both velocities. Another important observation is that constraint (10.3) is critical in determining the values of n and L, as these variables make it possible to define heat transfer surface area. Additionally, we know that at a constant heat transfer surface area, the objective function decreases when L is increased. This makes it easier to choose appropriate standardized double-pipe segments when assembling a heat transfer surface with the area equal to that calculated from constraint (10.3), especially if several standardized segments of different lengths are available.

The method consists of an exhaustive search in the decision space, checking consecutive values of variable m and of diameters d_1 and d_2; n and L are determined from constraint (10.3). The search begins at an initial solution approximation obtained on the basis of velocities of fluid flow in the exchangers as recommended in the literature. The logic diagram of the method is shown in Fig. 10.4.

In the present example, the initial approximation is obtained from the recommendation that toluene flow velocity should be $u_1 = 1.7$ m/s while for water $u_2 = 1$ m/s; consequently, cross-section areas of flow channels should be 0.384 m^2 and 0.401 m^2. Somewhat smaller cross-section areas (and thus, larger velocities) will be obtained in a heat exchanger assembled from $m = 3$ sections with pipe diameters $d_1 = 38$ mm and $d_2 = 57$ mm.

The calculation leading to the optimal solution is shown in Table 10.2. The first approximation, resulting from given flows of fluids combined with the initially assumed exchanger design, turns out to fall beyond the set of feasible solutions as the constraint on Δp_2 is violated. This indicates that flow velocity u_2 should be decreased, which can be accomplished by increasing number of sections m. At $m = 4$ both constraints are satisfied, indicating that the search direction has been correctly chosen. Subsequently, we check the numbers of sections $m = 5$ and $m = 6$. In the latter case the function value $f(x)$ increases: knowing that the function is unimodal, it would make no sense trying larger m values without changing the values of the remaining variables.

New pipe diameters are $d_1 = 57$ mm and $d_2 = 89$ mm, while a single section ($m = 1$) is enough to make the cross section areas close to their initial approximations. Constraints on Δp_1 and Δp_2, however, are violated for such an exchanger. At $m = 2$, both constraints are satisfied. In a subsequent point, at $m = 3$, the objective function increases, which again indicates that it is of no use increasing the number of sections even further.

As the assumption of larger pipe diameters caused the objective function to increase, it follows from the function unimodality that it is unnecessary to check double-pipe segments of even larger diameters. The search can thus be terminated. The optimal exchanger consists of $m = 5$ sections including the segments with pipes $d_1 = 38$ mm and $d_2 = 57$ mm; there are $n = 6$ elements of length $L = 4$ m in each section.

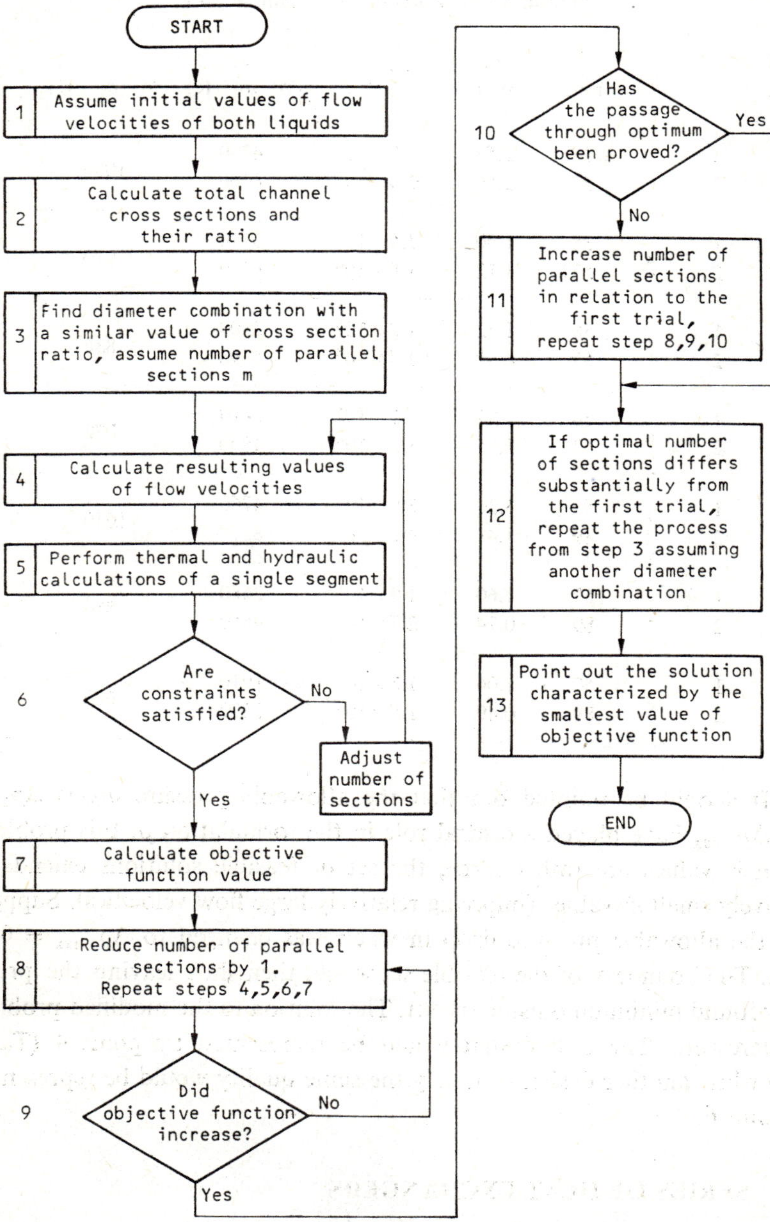

Fig. 10.4 Logic diagram of the calculations leading to the optimal choice of a double-pipe heat exchanger.

Table 10.2

Optimization procedure in Example 10.1

Point No.	i	m	d_i mm	u_i m/s	Re_i	α_i W/(m² · K)	k W/(m² · K)	F, m² calculated
1	1	3	38	2.53	2×10^5	4280	1000	11.4
	2		57	1.50	2.4×10^4	8700		
2	1	4	38	1.90	1.5×10^5	3370	910	12.5
	2		57	1.12	1.8×10^4	6850		
3	1	5	38	1.52	1.2×10^5	2800	836	13.6
	2		57	0.90	1.4×10^4	5690		
4	1	6	38	1.27	1×10^5	2410	772	14.8
	2		57	0.75	1.2×10^4	4890		
5	1	1	57	3.20	3.9×10^5	4760	1040	10.7
	2		89	1.48	4.6×10^4	8270		
6	1	2	57	1.60	1.9×10^5	2680	810	13.6
	2		89	0.74	2.3×10^4	4650		
7	1	3	57	1.06	1.3×10^5	1910	670	16.5
	2		89	0.49	1.5×10^4	3320		

It should be pointed out that the allowable pressure drops $\Delta p_{1\max}$ and $\Delta p_{2\max}$ have played a central role in the formulation of this problem. As their values are rather large, the set of feasible solutions extends to relatively small m values (implying relatively large flow velocities). Suppose that the allowable pressure drop in water was changed to $\Delta p_{2\max} = 0.02$ MPa. The boundary of the feasible set would then shift leaving the previously found minimum outside the set. The solution to the modified problem is interesting. The best design would be represented by point 4 (Table 10.2) while another design of nearly the same quality would be represented by point 6.

10.2 SERIES OF HEAT EXCHANGERS

In certain processing plants, heat exchangers constitute up to 25–40% of the total weight of the equipment. This indicates that considerable economic gains can be made by optimizing the heat exchanger design,

F, m^2 selected	L m	n	M kg	Δp_i MPa	Constraints satisfied?	$M(x)\varkappa_m(1/\tau)$ £	$C(x)$ £	$f(x)$ £
12.45	6	6	678	0.048 0.090	yes no			
13.36	3	10	769	0.030 0.050	yes yes	25.63	7.42	33.05
13.60	4	6	759	0.014 0.027	yes yes	25.30	3.85	29.15
16.02	3	8	920	0.010 0.019	yes yes	30.67	2.65	33.32
12.46	6	12	826	0.066 0.096	no no			
13.60	4	10	921	0.015 0.015	yes yes	30.70	2.95	33.65
18.69	6	6	1235	0.005 0.006	yes yes	41.16	1.06	42.22

manufacture, and operation. The design of a heat exchanger series is certainly a problem of economic importance.

The theoretical formulation of the problem of series design (Section 7.3) is entirely applicable to the heat exchanger series. Practical difficulties may arise, however, in obtaining input data necessary to utilize the theoretical model. Certain data needed for accurate optimization may be unavailable, as they are not recorded in routine industrial files or in production statistics. In such cases it may be necessary to modify the model, keeping its essential ideas intact, and making it possible to find an approximate solution.

As an example of series optimization on the basis of incomplete data, mention may be made of the design of a series of double-pipe heat exchangers [70]. The series had been designed for a manufacturer who introduced it in his production programme (Table 10.1). The fact that the data is incomplete has been caused by lack of a reliable economic analysis of the manufacturing process, i.e. no relationship could be established between the parameters of the series and manufacturing costs. An assumption has been made that all evaluations must rely on estimates of the exchanger operation, the effects of which evidently depend on the exchanger design and on the mode of operation. This implies an objective function $f(\boldsymbol{x}, \boldsymbol{y})$.

The introduction of additional constraints into the problem formulation can be avoided, as the rules of structural design can be entirely separated from the process requirements. For an arbitrary mode of operation, an optimal exchanger design $\boldsymbol{x}^{\mathrm{opt}}$ can be defined (see Fig. 7.1); this is understood to be a value $\boldsymbol{x}^{\mathrm{opt}}$ such that

$$f(\boldsymbol{x}^{\mathrm{opt}}, \boldsymbol{y}) = \min_{\boldsymbol{x} \in X} f(\boldsymbol{x}, \boldsymbol{y}) \tag{10.8}$$

where: X — space of feasible designs.

A satisfactory quality level of the series can be attained if a requirement is observed that for an arbitrary operation mode $\boldsymbol{y}$, a design $\boldsymbol{x}$ can be chosen from the series with such a property that the objective function value $f(\boldsymbol{x}, \boldsymbol{y})$ does not differ substantially from $f(\boldsymbol{x}^{\mathrm{opt}}, \boldsymbol{y})$. In other words, the series should be constructed in a way facilitating such a choice of exchanger design that the actual objective function value differs little from the lowest value which could be attained by a free design optimization.

The above postulate can be formally defined, if we assume the value of the allowable relative deviation γ of objective function value from its value at optimum:

$$f(\boldsymbol{x}, \boldsymbol{y}) \leqslant (1+\gamma) f(\boldsymbol{x}^{\mathrm{opt}}, \boldsymbol{y}) \tag{10.9}$$

Now, we define as an optimal series $\Gamma_\gamma = \{\boldsymbol{x}_i;\, i = 1, 2, \ldots, v_\gamma\}$ such a subset of space X in which for arbitrary $\boldsymbol{y}$, all elements satisfy postulate (10.9) and which consists of a minimal number of elements, i.e.

$$v_\gamma = \min \tag{10.10}$$

The problem stated above can be solved by the numerical methods of nonlinear programming aided by an auxiliary graphical technique.

Concerning the choice of the value of coefficient γ, no general recommendation can be given. In any specific series optimization problem, however, an estimate can easily be given of an acceptable value of γ. Moreover, such an estimate can be verified by checking on the value of v_γ (number of exchangers in the series) which must be reasonably large. From the practical point of view, the decision making model presented above always yields a series design which may be non-optimal but which certainly is good, provided only that the objective function has been wisely chosen. In checking an arbitrary series proposal, using the model can reveal exchanger designs covering overlapping regions of applicability. Also, gaps in the covering of the region of operation modes known from demand analysis can be discovered.

A segment of a double-pipe heat exchanger, shown in Fig. 10.3, can be uniquely defined by three numbers: pipe diameters d_1 and d_2 together with segment length L. One design characterized by three numbers can be interpreted as point $\boldsymbol{x}$ in three-dimensional space R^3. When designing the series, it was necessary to follow the standards mandatory on diameter combinations, as well as manufacturer's recommendations on segment lengths (length, in meters, should be an integer less than 6). In space R^3, decision space X was thus defined, as a finite set of points. The series can be considered as a subset of space X.

The operation of a double-pipe segment is essentially the pumping of liquids through segment channels. This is associated with the heat transfer across the wall of the inner pipe, induced by a temperature difference between the liquids. Each particular exchanger size can be operated in various modes. A complete definition of the operation mode requires 12 quantities to be specified: mass flows of both liquids G_1 nad G_2, temperatures at the extremes of the heat transfer surface t_1', t_1'', t_2' and t_2'', mean temperature difference Δt_m, combined thermal resistivity of fouling deposits at both surfaces of inner pipe r, surface film conductances at the same surfaces α_1 and α_2, heat transfer coefficient k and thermal capacity Q. This problem can also be treated as defining a point in a twelve-dimensional space; different operation modes are represented by different points in that space. If the physical properties of the liquids flowing through the exchanger are known, then the operation mode can be defined by only some of the quantities mentioned above. The relationships describing exchanger operation form a set of 7 equations. Heat balance implies the equations

$$G_1 C_{p1}(t_1' - t_1'') = Q \tag{10.11}$$

$$G_2 C_{p2}(t_2'' - t_2') = Q \tag{10.12}$$

where C_{p1}, C_{p2} – specific heats of liquids.

A fundamental design relationship, analogous to (10.3), is

$$\pi d_1 L k \Delta t_m = Q \tag{10.13}$$

The other equations are summarized below in symbolic form only as they have been discussed in greater detail in Section 10.1

$$\Delta t_m = f_m(t_1', t_1'', t_2', t_2'') \tag{10.14}$$

$$\alpha_1 = f_1(d_1, G_1, t_1', t_1'') \tag{10.15}$$

$$\alpha_2 = f_2(d_1, d_2, G_2, t_2', t_2'') \tag{10.16}$$

$$k = f_k(d_1, \alpha_1, \alpha_2, r) \tag{10.17}$$

where f_m, f_1, f_2 and f_k are known functions.

If a specific exchanger section design is considered (i.e. if d_1, d_2 and L are known) than the set of equations (10.11)–(10.17) relate seven coordinates of a relevant point in the twelve-dimensional space leaving five unspecified coordinates. Consequently, the operation mode can be defined in a five-dimensional space and the most convenient choice of coordinates is $G_1, G_2, t_1', \Delta t_m$ and r.

Now, it becomes possible to analyze operation modes for the exchanger segments in the production programme, even if no a priori information exists on the operation of every segment produced. Practical values of process quantities define operation space Y as a bounded region in a five-dimensional space: $t_1' = 0$ to 200°C, r between zero and values depending on the liquid type, Δt_m probably in the range 0 to 100°C while G_1 and G_2 depend on the properties of liquids.

The demand has been analyzed by investigating the orders that have been addressed to the manufactures during several years. The demand has been established to be uniformly distributed in space Y.

As an objective function for optimization of double-pipe segments, an index has been devised reflecting the cost of transferring a unit amount of heat

$$f(x, y) = \frac{\dfrac{M(x)\varkappa_m}{\tau} + C(x, y)}{Q(x, y)} \tag{10.18}$$

The symbols used in this formula are consistent with those in Section 10.1, but they refer to a single exchanger segment. Cost of power $C(x, y)$ consumed by the pumps is described by formula (10.7), with the same stipulation concerning the symbols.

As stated earlier, the operation mode can be investigated in the five-dimensional space of coordinates $G_1, G_2, t_2', \Delta t_m$ and r. The objective function exhibits monotone dependence on t_1', Δt_m and r so these quantities will not be considered here.

For each exchanger design at given t_1', Δt_m and r, the objective function depends on the values of G_1 and G_2 in the manner illustrated in Fig. 10.5. There exist mass flows G_1^{opt} and G_2^{opt} that minimize the function (this is the solution to problem type (2) discussed in Section 7.3). In a certain region in the neighbourhood of point (G_1^{opt}, G_2^{opt}) where function values do not deviate much from the minimal value f_0, the exchanger is operated in a nearly optimal manner. Such a region can be defined according to inequality (10.9) as a set of $\boldsymbol{y} \in Y$ satisfying the condition

$$f(\boldsymbol{x}, \boldsymbol{y}) \leqslant (1+\gamma)f_0 \tag{10.19}$$

where γ – allowable deviation of objective function value from its value at minimum.

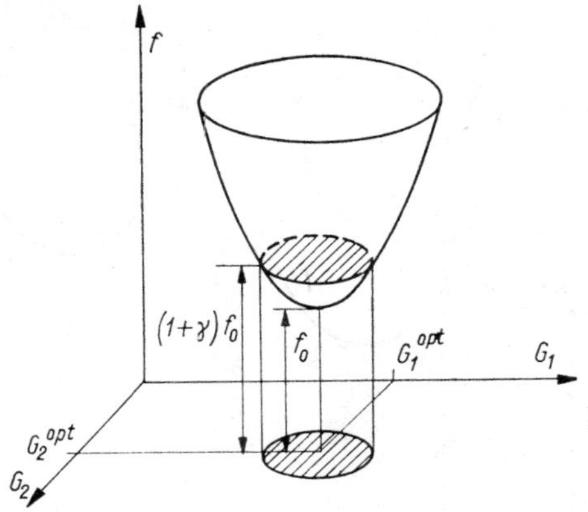

Fig. 10.5 The dependence of objective function f on mass flows of liquids. Shaded area on plane G_1, G_2 is the region of 'satisfactory' mass flow values.

If $\boldsymbol{x}$ (i.e. d_1, d_2 and L), t_1', Δt_m and r are given, then inequality (10.19) together with equations (10.11)–(10.17) permit determination of the boundary of the region of 'satisfactory' mass flows. To each exchanger design a region may be assigned on the G_1, G_2 plane that corresponds in space Y to a region of 'satisfactory' operation modes. According to condition (10.10), the optimal series should be reflected by a finite covering of space Y consisting of the lowest possible number of such regions.

The investigation of the covering of space Y can be transferred to a special and particularly convenient coordinate system. If design $\boldsymbol{x}$ is given, then for each pair of G_1 and G_2 values and known t_1', Δt_m and r – the remaining seven quantities (of the twelve that define the operation mode) can be determined by solving the system of equations (10.11)–(10.17). A particularly important quantity is thermal effect Q which is a measure of useful effect of exchanger operation. The covering can be thus visualized in coordinates f–Q. To various exchanger designs, various curves can be assigned in this coordinate system (Fig. 10.6). Conditions (10.9) and (10.10) can thus be replaced by a requirement that the chain of curve segments spanned over the projection of operation space on coordinate Q should consist of minimal number of segments positioned as low in the diagram as possible.

In the investigations of various series in terms of the scheme presented above, a combination of a graphical technique and the numerical methods of nonlinear programming has been successful. A computer program has been developed that, for any design and given t_1', Δt_m, r, computes

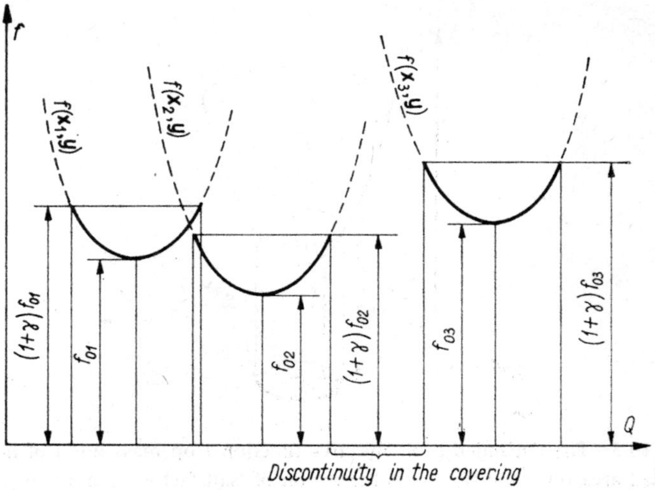

Fig. 10.6 Principle of recognizing the properties of a series of double-pipe segments. To designs x_1 and x_2, two neighbouring regions of satisfactory operation modes are assigned. There is a discontinuity in the covering of coordinate Q between the regions corresponding to designs x_2 and x_3.

optimal values of G_1, G_2 and then, for given γ, finds the boundaries of the region defined by inequality (10.19). Series characteristics in coordinates f–Q can subsequentlly be displayed graphically.

The starting point for the calculations was the check on the series design suggested by the manufacturer. It included 28 segment sizes (7 pipe diameter combinations and 4 length variants). Parameter values were as follows: equipment life time $\tau = 6$ years, pumping efficiency $\eta = 0.75$; power cost $\varkappa_e = 0.002$ £/kWh, annual utilization time $\theta = 8000$ h; cost of carbon steel tubes in the relevant diameter region $\varkappa_m = 0.2$ £/kg. Series characteristics were determined assuming various liquids in the exchanger channels. It has been proved that at $\gamma = 0.10$–0.12, continuity is ensured of the chain of regions of satisfactory operation.

In Fig. 10.7, an example of results is shown in coordinates f–Q, for one of the combinations of t_1', Δt_m and r (water flow in both channels is assumed). This graph can be interpreted as follows:

– the series according to the manufacturer's suggestions is a good starting point to series optimization because a finite covering of space Y has been attained;

– the number of exchanger sizes in the series is too large because for lower duties several sizes are available but they differ in quality.

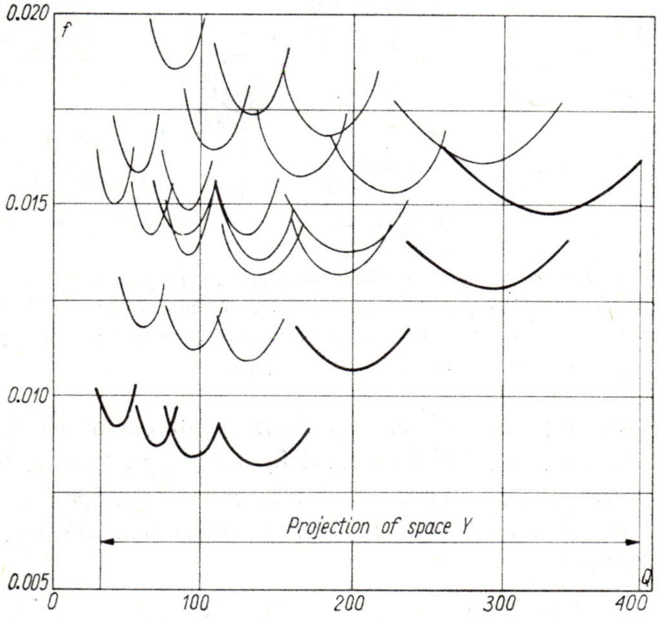

Fig. 10.7 Result of an analysis of a series of double-pipe segments including 28 sizes. Seven curve segments marked by thick lines form a covering of the projection of space Y on axis Q; 7 corresponding designs thus form a correct series.

If we turn our attention to the curve segments marked by heavy lines, we can see that these segments form a chain, demonstrating thereby that conditions (10.9) and (10.10) are satisfied. A finite covering of the operation space is thus attained when there are seven sizes in the series; this is 1/4 of the number initially suggested. At the same time, the number of diameter combinations is reduced to 4.

10.3 REFRIGERATION EQUIPMENT

In refrigeration systems (Fig. 10.8) specialized heat exchangers known as condensers and evaporators are widely used. The process requirements imposed on their design form a basis for various optimization problems. This can be exemplified by a problem related to evaporators in which the refrigerant boils at the expense of heat absorbed from the space cooled (for example, from a cold room or from the chamber of a refrigerator).

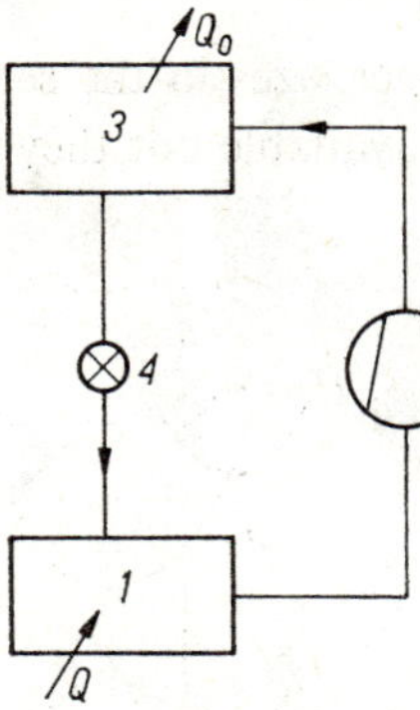

Fig. 10.8 Scheme of a simple refrigeration system. 1 – evaporator, 2 – compressor, 3 – condenser, 4 – throttling valve. Q_0 and Q – thermal effects, respectively: dissipated to the environment and absorbed from the cooled space (Q is also known as refrigerating effect).

The analysis presented below is concerned with one of the most widely used evaporator designs characterized by refrigerant flow in horizontal tubes. In principle, this is a horizontal, tube-in-shell, multipass heat exchanger. Its cost depends on the heat transfer surface area which in turn can be expressed as

$$F = \frac{Q}{k\Delta t} \tag{10.20}$$

where: Q – refrigerating effect, k – heat transfer coefficient, Δt – mean temperature difference.

It would seem that advantageous designs could be obtained, if a sufficiently high refrigerant flow velocity is used producing large heat transfer coefficients. Increased flow velocity, however, results in increased hydraulic loss which in turn causes the boiling temperature to decrease in the direction of flow. Consequently, as the required temperature in the cooled space is constant (it is a parameter in the design problem), the mean temperature difference decreases. This counteracts the advantageous effects which could have been obtained by the increased heat transfer coefficient.

In conclusion, an increase in refrigerant flow velocity makes sense only if the increase of heat transfer coefficient is not offset by the decrease of mean temperature difference.

In practical design problems, the advantages offered by increased heat transfer and the disadvantages of reduced temperature difference depend on the proportioning of the refrigerant flow path in the evaporator. The most important design factors are tube diameter, number of tubes and number of passes. A design decision can always be made which represents a trade-off characterized by minimal heat transfer surface area.

In a typical problem of evaporator design, the decision variables are:
– tube diameter,
– number of passes.

The parameters are as follows:
– mass flow of the cooled medium,
– inlet and outlet temperature of the cooled medium,
– final refrigerant temperature at the evaporator outlet.

In the mathematical model, design requirements can be included, for example those resulting from the usual limitations on evaporator dimensions.

The most complicated part of the problem is the identification of the objective function. This can be heat transfer surface area (according to formula (10.20)), evaporator weight or evaporator cost. In a detailed formulation, it is necessary to rely on algorithms for computation of heat transfer coefficient and mean temperature difference as functions of the velocity of refrigerant flow inside the tubes. In calculating mean temperature difference, it is necessary to know how the hydraulic loss depends on refrigerant flow velocity. As a rule, this relation must be experimentally verified because, in the literature, appropriate formulae can only be found for certain refrigerants [11]. Correct problem formulation depends on the possibility of determining local values of surface film conductance in the refrigerant because this is required in order to account for the coupling of heat transfer and flow phenomena.

Different methods of solving the problem of optimization of a tubular refrigerating evaporator can be found in the literature [29, 75]. All these methods have been formulated on the basis of practical process knowledge, rather than from an analysis of formal properties of the corresponding mathematical model. In [75] where many simplifying assumptions have been made, a suggestion is advanced that the calculation of heat transfer surface area should be carried out for each value of the number of passes, n. Each time, tube diameter is determined for the most advantageous refrigerant flow velocity, and thus for minimal heat transfer surface area. The choice of the final solution results from a comparison of the designs obtained. An example of such a comparison is shown in Fig. 10.9. The graph makes it possible to analyze solution sensitivity to deviations from the optimum. The absolute optimum is obtained when the number of passes, $n = 5$, while the solution characterized by $n = 4$ is only a little worse. The designer is thus free to choose between a solution characterized by the positions of refrigerant inlet and outlet at the same evaporator end (even number of passes), and a solution with inlet and outlet positions at opposite ends (odd number of passes).

Another problem originating from refrigeration technology and also encountered in thermal power generation, is the optimization of air coolers. Such units are frequently used as condensers in refrigeration systems. Their task is to dissipate, to the environment, the total heat which is the sum of that removed by refrigeration and that generated by the compressor.

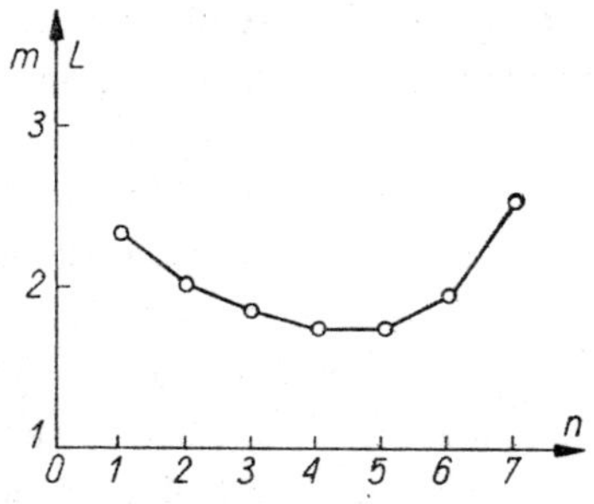

Fig. 10.9 Example of a relationship between total length of evaporator tubes and number of passes.

A characteristic feature of air cooler design is that the heat transfer surfaces consist of tubes, usually finned. The geometry of the tubing, including that of the fin, may be chosen by the designer.

An important point in the air cooler operation is that the power consumption of the cooling fan is much greater than that of the refrigerant pump. As a consequence, where economic considerations govern cooler design, the cost of refrigerant pumping may be neglected.

The above properties of air coolers form the basis on which optimization problems are formulated. Their aim is to choose properly air flow conditions in the cooler so that the overall cost is minimal. A variety of problems of this kind can be found in the literature [25, 42, 58]. The problems differ with respect to the number of design details included, from determination of air flow velocity, or Reynolds number, to detailed proportioning of cooler tubing.

Among the simplest uses of the above approach is the problem reported in [25], which is concerned with the most economical air flow velocity in a cooler assembled from bare tubes. Velocity u is a decision variable, while the parameters are: volume flow of air V, thermal effect of the condenser Q, tube diameter d, tube pitch e, and several economic quantities mentioned below in the problem statement.

No constraints are specified. The total annual cost is adopted as the objective function:

$$f(u) = a\varkappa_F F + C_e$$

where: a – fixed cost coefficient, $\varkappa_F$ – unit cost of the cooler (per 1 m^2 of heat transfer surface area), F – heat transfer surface area, C_e – annual operation cost.

Heat transfer surface area is a function of air flow velocity:

$$F = \frac{Q}{k(u)\Delta t}$$

and the relation between heat transfer coefficient k and velocity u is

$$k(u) = \frac{1}{A + \dfrac{1}{\alpha(u)}}$$

where: A – a constant reflecting heat transfer on the refrigerant side and across the tube wall, α – surface film conductance at the outer surface of the tubes.

The conductance α depends on air flow velocity

$$\alpha(u) = Bu^{0.61}$$

where B is a constant.

Annual operation cost can be approximately calculated as

$$C_e = \frac{\Delta p V}{\eta 10^3}\varkappa_e \theta$$

where: $\varkappa_e$ – energy cost in £/kWh; θ – annual utilization time of the cooler, h; η – fan efficiency.

The pressure drop accompanying air flow in the cooler:

$$\Delta p = Cu^2$$

where C is a constant. For a given tubing configuration, its value can be found in the literature [25].

By substituting the above relations into the annual cost formula, we obtain:

$$f(u) = a\varkappa_F \frac{Q\left(\dfrac{1}{Bu^{0.61}} + A\right)}{\Delta t} + \frac{Cu^2 V\varkappa_e \theta}{\eta 10^3}$$

With increasing velocity u, the first component (fixed cost) decreases while the second component (operation cost) increases. It is thus reasonable to expect an optimum resulting from a trade-off between these conflicting tendencies. Upon differentiation of the above function with respect to variable u, we establish the optimal air velocity

$$u^{\text{opt}} = \left(\frac{305 a\varkappa_F Q\eta}{\Delta t BCV\varkappa_e \theta}\right)^{0.383}$$

Optimization of Reynolds number for the air flow in coolers equipped with finned tubes has been similarly solved in [42]. The problem has been formulated more generally than in the previous case using the definition of Reynolds number

$$Re = u\frac{d\rho}{\mu}$$

where: d – diameter of bare tube (without fins); ρ and μ – density and absolute viscosity of air, respectively.

As a consequence, the result of the optimization sets up a basis for simultaneous choice of air velocity and tubing geometry.

The parameters in this problem are economic data which follow and coefficient $\varkappa_t$ reflecting temperature relations in the heat transfer process ($\varkappa_t$ is also explained below). There are no constraints in this problem. Total annual cost has been adopted as an objective function:

$$f(Re) = C_1 + C_2 + C_3 \tag{10.21}$$

where C_1, C_2 and C_3 are the fixed cost, the cost of power consumed by the air fan and the maintenance cost.

The components of sum (10.21) have been determined as follows:

$$C_1 = a\varkappa_F F(Re) \tag{10.22}$$

where: a – coefficient of fixed cost, $\varkappa_F$ – cost of 1 m² of heat transfer surface; F – heat transfer surface area;

$$C_2 = b\,\frac{G\Delta p(Re)\varkappa_e}{\rho\eta 10^3}\,\theta \tag{10.23}$$

where: b – fan load factor, G – mass flow of air, Δp – fan pressure increase, η – fan efficiency, $\varkappa_e$ – power cost, £/kWh, θ – annual cooler utilization time, hours

$$C_3 = c\varkappa_F F(Re) \tag{10.24}$$

where: c – coefficient of maintenance cost.

The minimum of function (10.21) stems from the same reasons as in the previous example: with increasing Re, $C_1 + C_3$ decreases, while C_2 increases. Minimal cost is found by introducing (10.22)–(10.24) into (10.21) combined with a simultaneous account of detailed relations between heat transfer surface area and pressure head on the one hand and Reynolds number on the other. The new function is then differentiated with respect to Re and, as a result, a relation is obtained between the optimal Reynolds number and the parameters:

$$Re^{\text{opt}} = \left(\frac{10^3 \eta}{\theta} Y \frac{\varkappa_F}{z}\right)^{\frac{1}{2+B-D}} \tag{10.25}$$

where: z – number of tube rows, B and D – constants, and Y is defined by a formula

$$Y = \frac{a+c}{b} \frac{\varkappa_t}{\varkappa_e} \frac{Bd^3}{\frac{F_z}{F} AE(2-D)} \frac{\rho^2 C_p}{\mu^2 \lambda}$$

where: A, E – constants, F_z/F – ratio of outer tube surface area to inner tube surface area, λ – thermal conductivity of air; C_p – specific heat of air.

Coefficient $\varkappa_t$ reflects the conditions of the heat transfer process:

$$\varkappa_t = -\ln\left(1 - \frac{\Delta t_p}{t - t_p}\right)$$

where: Δt_p – air temperature increase in the cooler, t – refrigerant temperature, t_p – final air temperature. At optimal process conditions, $\varkappa_t = 1.255$.

Actually, the presentation of the solution to this problem given in [42] is not entirely correct. When calculating the value of expression (10.25), the author accounts for some scatter of values of coefficients, $\varkappa_F$, A and B, given in the literature. Thus, instead of a unique relation between optimal Reynolds number and the number of tube rows, a vague recommendation is given. In Fig. 10.10, recommended Reynolds number values are given at a fan efficiency $\eta = 0.65$, annual cooler utilization time $\theta = 8000$ h, fan load factor $b = 0.525$ and economic data which can be regarded as typical for air coolers. Actually, the scatter of values of coefficients should

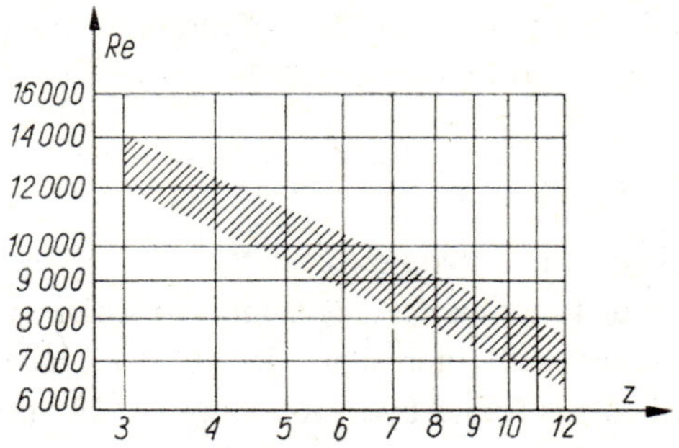

Fig. 10.10 Optimal Reynolds number in air flow in a cooler featuring finned tubes, as a function of the number of tube rows.

be interpreted as incomplete information. This could then lead to an appropriate form of rigorous presentation of the results.

The applications of optimization methods to air cooler structure design are extensively treated in [58]. Because of limited space this subject cannot be covered here. An idea of the difficulty of structure design problems may be obtained from Table 10.3, where possible decision variables are listed together with some additional information. Reference [58] also contains a discussion of mathematical models and objective functions, a survey of the methods of solution, and two examples of practical designs.

Table 10.3

Decision variables in typical design problems related to air coolers

Name	Type	Number of values allowed by standards
Tube length	discrete	about 10
Bundle width	discrete	about 15
No. of tube rows	integer	up to 10
Tube diameter	discrete	1–2
Pitch	discrete	2–3
No. of passes	integer	up to 9
Mass flow	continuous	
Fin spacing	discrete	up to 6
Fin height	discrete	up to 4
Fan diameter	discrete	10–15
No. of blades in fan rotor	integer	2–3

CHAPTER 11

Miscellaneous equipment

11.1 GEAR PUMP

Typical gear pump applications are characterized by a relatively high pressure (5–15 MPa) and a small volume flow (several litres per second). Owing to their structural simplicity and compactness, as well as to easy operation and maintenance, gear pumps find widespread use in various technical systems, for example in hydraulic drives and hydraulic control systems.

A gear pump, Fig. 11.1, consists of a casing and two gear wheels which are mounted in the casing with a minimum clearance. Liquid from the suction space of the pump (space 1) enters tooth spaces of both gear wheels. Wheel rotation causes liquid movement to the delivery space of the pump (space 2).

Theoretical gear pump capacity depends on gear wheel dimensions, tooth parameters and rotational speed. The effective capacity is less than the theoretical, the ratio being the volumetric efficiency of the pump with

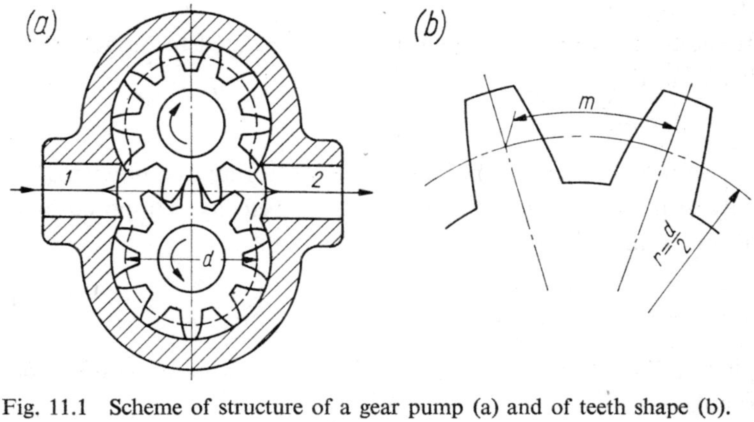

Fig. 11.1 Scheme of structure of a gear pump (a) and of teeth shape (b).

a value usually between 0.9 and 0.95. Designing a gear pump involves, in fact, a trade-off between possibly high operational merits (above all, compactness) and unavoidable manufacturing requirements related to tooth parameters and gear wheel life. All these aspects can be reflected in the mathematical model and in the objective function.

Decision variables are:
- number of teeth, z
- wheel width b,
- module pitch m,
- rotational velocity n.

Parameters are:
- pump capacity V,
- forcing pressure p,
- characteristic quantities of operational limitations, i.e. maximum and minimum allowable linear velocity at pitch diameter of gears c_{max}, c_{min} and volumetric efficiency η_v,
- characteristic quantities of manufacturing limitations, i.e. minimum and maximum number of teeth z_{min}, z_{max}; minimum and maximum module pitch m_{min}, m_{max}; minimum and maximum ratio of wheel width to module pitch B_{min}, B_{max}; nominal pressure angle α_0.

The mathematical model is based on a few fundamental relations. Pump capacity as a function of design quantities [73]

$$V = 2\pi Bnm^2(z+\sin^2\alpha_0)\eta_v \tag{11.1}$$

Peripheral velocity at pitch diameter of gears

$$u = \frac{\pi dn}{60} \tag{11.2}$$

where: d – pitch diameter.

It should be noted that pitch diameter is related to module pitch and number of teeth

$$d = mz \tag{11.3}$$

When formulating operational and manufacturing requirements, it can be easily seen that the limitation of the linear velocity is imposed by a constraint on the product nmz

$$c_{min} \leqslant nmz\frac{\pi}{60} \leqslant c_{max} \tag{11.4}$$

By combining relations (11.1) and (11.2), a constraint on product bm can be formulated

$$bm \geqslant \frac{V}{0.1\eta_v c_{max}} \tag{11.5}$$

(this formula includes the approximation $2\pi \approx 6$). The limitations on gear wheel width can thus be expressed as

$$\frac{B_{min}}{c_{min}} \leqslant b^2 \frac{0.1\eta_v}{V} \leqslant \frac{B_{max}}{c_{max}} \tag{11.6}$$

From formula (11.2), limitations of rotational velocity of wheels are derived

$$\frac{60c_{min}}{\pi d} \leqslant n \leqslant \frac{60c_{max}}{\pi d} \tag{11.7}$$

The mathematical model is now complete. Now we can attempt to find a measure of the most important design feature of a gear pump, that is, its compactness. This measure will be helpful in expressing the compactness as an objective function which depends on the decision variables. Now, this requirement can be satisfied by a simple expression which has the meaning of longitudinal cross section area of one gear wheel

$$f^*(b, m, z) = bmz \tag{11.8}$$

Constraints (11.4)–(11.7) and objective function (11.8) (which should be minimized) represent the problem of designing the most compact gear pump arrangement.

The solution procedure can be substantially simplified by taking advantage of the properties of the expressions containing the decision variables. By taking their logarithms and introducing new variables, all products of variables can be transformed into sums and thus the problem becomes a linear one. Accordingly, upon taking logarithms of constraints and objective function and upon substituting

$$x_1 = \ln z, \qquad x_2 = \ln b, \qquad x_3 = \ln m, \qquad x_4 = \ln n$$

we obtain the following problem

$$x_1+x_3+x_4 \geqslant a_1, \qquad x_1+x_3+x_4 \leqslant a_2$$

$$x_1 \geqslant a_3, \qquad x_1 \leqslant a_4$$

$$x_2+x_3 \geqslant a_5, \qquad x_2 \geqslant a_6, \qquad x_2 \leqslant a_7$$

$$x_3 \geqslant a_8, \qquad x_3 \leqslant a_9$$

$$f(x_1, x_2, x_3) = x_1+x_2+x_3 \rightarrow \min$$

where $a_1, a_2, \ldots, a_9$ are numbers obtained upon transformation of free terms in the original constraints, for example

$$a_1 = \ln \frac{60c_{min}}{\pi}, \qquad a_3 = \ln z_{min}, \qquad \text{etc.}$$

A characteristic feature of the problem above is that the objective function is independent of decision variable x_4. Nevertheless, this variable appears in the constraints and thus the choice of its value is as important as the choice of values of the remaining variables.

The problem of optimization of a gear pump has been solved in [73] for the following parameter values: $V = 1.4$ dm^3/s, $p = 10$ MPa, $\eta_v = 0.96$, $c_{max} = 5.7$ m/s, $c_{min} = 4.5$ m/s, $z_{min} = 8$, $z_{max} = 14$, $B_{min} = 6$, $B_{max} = 10$, $m_{min} = 0.002$ m, $m_{max} = 0.006$ m. After determining parameters a_1–a_9 in the constraints, the resulting linear programming problem has been solved by the modified simplex method. The solution is $x^{opt} = (1.19, 3.39, 1.61, 7.55)$ and the corresponding values of original decision variables are: $z^{opt} = 8$, $b^{opt} = 0.02978$ m, $m^{opt} = 0.0055$ m, $n = 2000$ rpm.

11.2 MULTIPLE HYDROCYCLONE SYSTEM

A multiple hydrocyclone system is a device to separate dispersed material from a liquid by means of centrifugal action. It consists of a train of parallel connected identical hydrocyclones, each of them shaped as a cylindrical shell closed at the top by a flat plate and ending at the bottom in a cone with an opening at its apex (Fig. 11.2). Fixed in the centre of the top plate is a cylindrical tube. The liquid or suspension to be separated is injected tangentially into the upper part of the cylindrical section, causing the liquid contained in the cyclone to rotate with a resultant increased concentration of material particles:

– at cyclone wall, if solid material density is greater than liquid density,

– in the vicinity of cyclone axis, if the relation between the densities is reversed.

In the first case, the suspension with increased concentration of solid material is discharged through the outlet at the apex, while the clarified liquid is discharged through the central outlet on the end of the cyclone. In the second case, the roles of outlet nozzles are reversed. For a complete design of a multiple hydrocyclone system, it is necessary to define the hydrocyclone geometry and the number of cyclones to make up the system. Recommendations on the proportioning of hydrocyclones can be found in the literature [66]. It is also known from experiment that the most advantageous separation effect is attained at the minimum value of the characteristic number

$$C_{y50} = \frac{d_{50}^2 \Delta\rho}{\mu} \frac{L \Delta p}{\rho q} \tag{11.9}$$

where: d_{50} – diameter of particles separated to 50%, $\Delta\rho$ – difference between density of solid and that of liquid, μ – absolute liquid viscosity, L – total length of cyclone from top plate to apex, Δp – static pressure drop from inlet to outlet, ρ – liquid density, q – throughput of cyclone.

According to Rietema's experiments [66], C_{y50} takes a minimum value (equal to 3.5) for cyclone geometry defined as follows:

$$\frac{L}{D} = 5, \qquad \frac{d_1}{D} = 0.28 \tag{11.10}$$

$$\frac{d_2}{D} = 0.34, \qquad \frac{l}{D} = 0.4$$

Assuming this configuration of hydrocyclone as suitable, the designing of a multiple hydrocyclone system is reduced to the determination of the number n and diameter D of cyclones.

The essential design idea is to define multiple hydrocyclone system

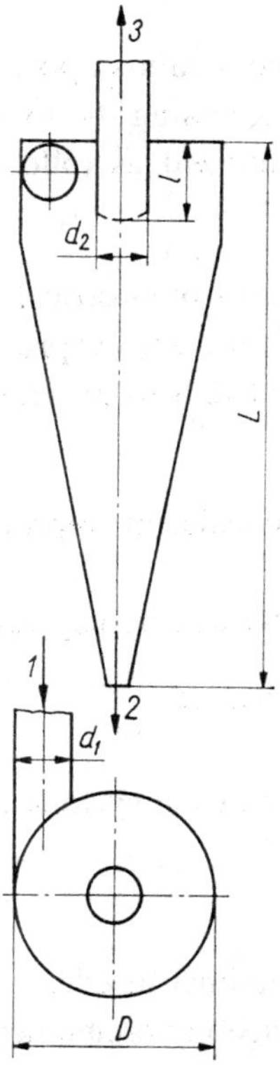

Fig. 11.2 Design principle of a hydrocyclone. 1 – inlet, 2 – underflow, 3 – overflow.

in a manner facilitating a trade-off between fixed cost and running cost. If the active volume of the unit is large (many cyclones, possibly of large dimensions), then the pressure loss is small and so is the cost of power consumed in pumping the suspension. On the contrary, at a small volume of the unit, a considerable pressure loss cannot be avoided. The first case is thus characterized by a large capital cost and a small running cost, while in the second case, the relation is reversed.

The decision variables in a multiple hydrocyclone system design are:

– number of cyclones m,

– cyclone diameter D,

and the parameters are as follows:

– suspension flow V,

– characteristic particle diameter d_{50},

– properties of solid material and liquid: ρ, μ, $\Delta\rho$,

– economic quantities, defined below in the explanation of the objective function, such as unit energy cost $\varkappa_e$, unit value of the recovered liquid $\varkappa_r$, etc.

In the general case, no constraints appear in the design problem.

The objective function expressing the total cost of the multiple hydrocyclone system can be summarized as follows

$$f(\boldsymbol{x}) = C_1 + C_2 - W \tag{11.11}$$

where: C_1 – fixed cost, C_2 – cost of electrical energy consumed, W – value of the recovered liquid (all components expressed per second of operation).

The fixed cost is reported [20] to depend on the cyclone diameter:

$$C_1 = n a_1 a_2 D^b \tag{11.12}$$

where: a_1 – constant, a_2 – coefficient representing capital depreciation, b – constant.

Pressure loss in a cyclone can be expressed by an empirical formula

$$\Delta p = \frac{\rho u^2}{2} 0.093\, Re^{0.3748} \tag{11.13}$$

where Re – Reynolds number at the cyclone inlet

$$Re = \frac{\rho u d_1}{\mu} \tag{11.14}$$

where: u – flow velocity in the inlet nozzle.

Remembering that suspension volume flow through a single cyclone is

$$q = \frac{V}{n}$$

we can express flow velocity as

$$u = \frac{4V}{n\pi d_1^2} \tag{11.15}$$

The power cost

$$C_2 = V\Delta p \varkappa_e$$

can be expressed with the aid of relations (11.13)–(11.15) as

$$C_2 = 0.093 V \varkappa_e \frac{\rho}{2} \left(\frac{0.28 D\rho}{\mu} \right)^{0.3748} \left(\frac{4V}{n\pi 0.28^2 D^2} \right)^{2.3748} \tag{11.16}$$

and the value of the recovered liquid is

$$W = V\varkappa_r \tag{11.17}$$

By combining equations (11.9), (11.10), (11.13) and (11.15), the following relation is obtained

$$D = \left[108.2 \frac{d_{50}^2 \Delta\rho}{C_{y50}\mu} \left(\frac{\rho}{\mu} \right)^{0.3748} V^{1.3748} \right]^{0.2963} n^{0.4074} \tag{11.18}$$

Formulae (11.11), (11.12), (11.16), (11.17) and (11.18) enable the cost to be expressed as a function of one variable n. The minimum of this function can be found analytically, although a graphical technique can be recommended for design applications. This technique is particularly suitable to investigate the objective function sensitivity to deviations of variable n from the optimum.

In Fig. 11.3 is shown an example of a graphical solution to the problem of the optimization of a multiple hydrocyclone system, given the following

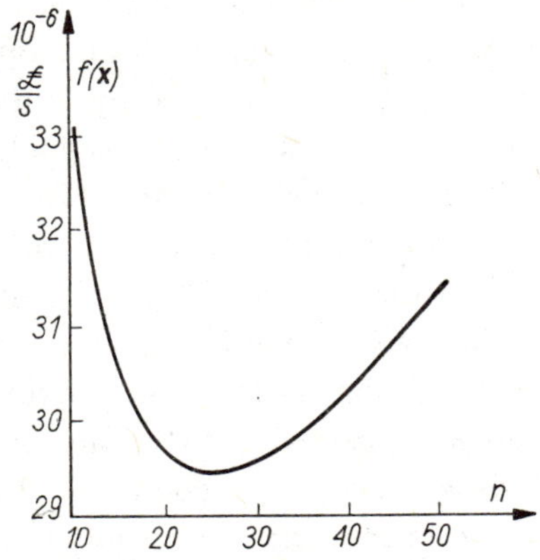

Fig. 11.3 Result of optimization of the number of cyclones in a multiple hydrocyclone system ($f(x)$ is the cost per second of operation).

input data: $V = 0.0155$ m^3/s, $d_{50} = 4 \times 10^{-6}$ m, $\mu = 0.0007$ kg/(m · s), $\rho = 760$ kg/m^3, $\Delta\rho = 1040$ kg/m^3, $\varkappa_r = 0$ and the remaining economic coefficients characteristic of British industry [20]: $a_1 =$ £ 1900 and $b = 1.26$ (for D expressed in metres in formula (11.12)), $a_2 = 3.18 \times 10^{-8}$ s^{-1}, $\varkappa_e =$ £ 1.70×10^{-9}/J. The cost curve attains its minimum at $n = 25.55$; accordingly, the optimal number of cyclones is $n_0 = 26$. The corresponding optimal value of cyclone diameter, calculated from (11.18), is $D_0 = 0.0268$ m. The shape of the cost curve in the vicinity of the optimum indicates, however, that it would be possible to adopt a number of cyclones different from 26 without substantially affecting design quality. This conclusion may be important because a standard design practice is to connect hydrocyclones into banks containing four cyclones each. It would thus be convenient to choose a number of cyclones which can be divided by four: either 24 hydrocyclones (diameter 0.0259 m) or 28 hydrocyclones (diameter 0.0276 m). When making the final decision, we can account for the fact that the first design has lower capital cost, while the second can easily absorb a suspension flow somewhat larger than required, which may contribute to the ease of system operation. Depending on requirements imposed by the process plant, one of these factors may prevail.

11.3 FIRED HEATER

Fired heaters are among the most important equipment in many process plants, for example in petroleum refineries. A fired heater is a heat exchanger in which a liquid flowing in a system of tubes raises its temperature by absorbing heat from hot gases generated in the combustion chamber. Usually, the heating is aimed at ensuring proper conditions for unit operations performed in subsequent stages of the process.

Fired heaters are difficult to manufacture and operate and are also expensive. Many structural elements, such as pipes in the heating surface, return bends, pipe hangers, chamber walls and gas guide vanes, are made of heat-resistant steel. Heating surface elements must be made of steel grades resistant to hydrogen corrosion. Steel elements account for 40% of the total heater weight; 1/3 in the form of heat resistant pipes and 1/20 as heat resistant castings which are responsible for the major part of the cost of materials. Under such circumstances, great importance must be attributed to design decisions concerned with economical proportioning of the heating surface. It is necessary to account for the particular process requirements and also for manufacturer's assumptions concerning standardization of pipes, return bends, supporting structures and heater instrumentation.

There are several modifications of the fired heater design. In modern oil refineries, heaters with cylindrical combustion chambers are most frequently used. In Fig. 11.4, two different structural principles for fired heaters are shown. The walls of the combustion chamber are covered by pipes heated mainly by radiation. This part of the heating surface is called the radiation section. Hot gases leaving the combustion chamber enter the convection section, where a transversely positioned tube bundle is heated mainly by convective heat transfer. The cold liquid enters the pipes in the convection section and then passes to the radiation section where it attains its final temperature. Bare heating pipes are usually employed in the radiation section, while in the convection section, studded or finned pipes can be also encountered.

The design of the heating surface of a fired heater is directed to such sizing of the surface elements that, at a specific flow rate of liquid, the required outlet temperatures are attained. Correct design must also ensure

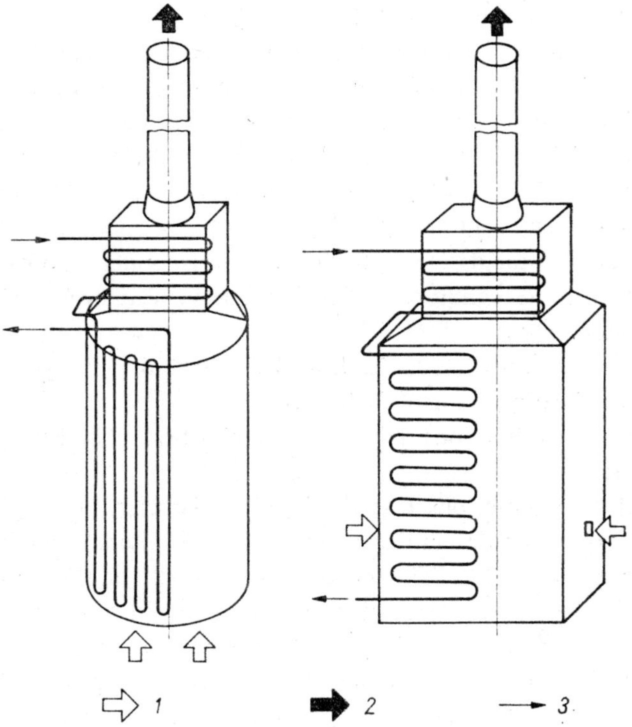

Fig. 11.4 Principle of proportioning heating surfaces in fired heaters as coils corresponding to single flow paths. 1 – air and fuel, 2 – flue gas, 3 – heated liquid.

that the heat flux on the pipe surface in the radiation section is contained within allowable limits, or that free thermal expansion of the pipes is possible.

As the design calculations are rather complicated and there is a variety of individual operational and manufacturing requirements, the traditional design technique consists of calculations and decision making combined with the production of drawings of the heater structure. The design may be assessed from the drawings immediately and, if the result of the evaluation is negative, then decisions can be modified and the design repeated until a satisfactory solution is found.

In order to introduce an optimal approach, it has been necessary to re-formulate the design procedure into a form which can be regarded as an application of a mathematical model. As there are many complicated relations, the model has been transformed into an algorithm subsequently coded as a computer program. Given the necessary input data, this program automatically produces results which enable the designer to make drawings of the heating surface [54, 69].

Owing to lack of space, the mathematical model of a fired heater cannot be presented here. It is worth mentioning, however, that certain questions arising in the formulation seem to be characteristic of the design of equipment described by a large number of process and design variables. Among these questions, the choice of the decision variables and objective function deserves particular attention.

Input data to the design algorithm can be separated into four groups:

- data on site conditions (elevation above sea level, ambient temperature, etc.);
- physical properties of the heated liquid and fuel (specific heat of liquid, heating value of fuel, etc.);
- general assumptions (shape of combustion chamber, pipe type, etc.);
- specific assumptions (excess air, pipe diameter, etc.).

The quality of heater design depends on the assumptions made in the third and fourth groups of data. The practical possibilities for their optimization are different. Most of the general assumptions consist of choosing between a few possible decisions and their consequences may be complex. For instance, the shape of the combustion chamber has a profound effect on the problems of fabrication, transportation and assembly of the heater structure. The consequences of the decision on the combustion chamber shape could perhaps be traced and economically evaluated, but this cannot be done at the design stage. For a particular application, chamber shape should be decided in accordance with the manufacturer's policy, rather than on the basis of optimization results.

The case of specific assumptions is different. Typically, a certain value should be chosen from a set of values defined by technological conditions. For instance, flue gas velocity in the convection section should be chosen from a continuous range of values characteristic of gas-heated exchangers. Similarly, pipe diameter must belong to a discrete set of values defined by relevant standards.

Under such circumstances, decision variables should preferably be chosen among quantities in the group of specific assumptions. A detailed analysis reveals the following quantities to be considered:

- share of radiation section in the total thermal effect, ψ,
- thermal effect of combustion per unit cross section area of the combustion chamber, q_A;
- allowable liquid velocity in tubes, c_r;
- allowable flue gas velocity in convection section, c_c;
- excess air, λ;
- tube diameter in radiation section, d_r;
- tube diameter in convection section, d_c;
- distance between the tubes in radiation section, e_r;
- distance between the tubes in convection section, e_c;
- flue gas velocity in stack, c_s.

For practical reasons, some of these quantities are not likely to be optimized. Typically, tube diameter and spacing are assumed to be the same in both sections: $d_r = d_c = d$ and $e_r = e_c = e$. Moreover, the flue gas velocity in the stack is not necessary for sizing the heating surfaces in the two heater sections.

Similarly, the formulation of the objective function for a fired heater is far from obvious. Qualitative design objectives may vary according to the circumstances and the requirements may also be different. Consequently, it may become necessary to formulate various objective functions and constraints. The mathematical model, then, should be combined with a carefully selected optimization method. Such a calculation tool could be later applied to various practical problems by changing the mathematical forms of the objective function and constraints.

The figures reflecting qualitative design objectives should be related to the economic consequences of design decisions. The objective function should thus include the capital and running costs of the heater. Such a function would be nonlinear, because the relation between the weight (and the cost) of the heating surface and geometrical variables (diameters, distances) is also nonlinear. Furthermore, the objective function is discontinuous with respect to these variables, because design algorithms include the rounding off of values to the nearest integer or to the nearest even number. A numerical method for design optimization must be insensitive to such function properties.

For exactly this reason, a '*crude*' *Monte Carlo method* has been selected for the optimization of fired heaters. A typical application is a problem featuring a vector of seven decision variables (identical diameters and spacings of the pipes have been assumed for both sections and gas velocity in the stack has been excluded from consideration).

The heating surface design defined by vector $\boldsymbol{x}$ was evaluated using the total weight $f(\boldsymbol{x})$ of pipes and connecting elements (elbows and return bends). The design characterized by minimum weight was considered as optimal.

In the constraint system, three inequalities were included relating three quantities to their largest allowable values $q_i^{\max}$, $i = 1, 2, 3$, namely:

- pressure drop in the liquid flowing through the heater, $q_1(\boldsymbol{x})$;
- average heat flux at the pipe surface in the radiation section, $q_2(\boldsymbol{x})$;
- thermal effect of fuel combustion per unit volume of the combustion chamber, $q_3(\boldsymbol{x})$.

Introducing the notation

$$q_i(\boldsymbol{x}) - q_i^{\max} = h_i(\boldsymbol{x})$$

the problem of heater optimization was formulated as determine $\boldsymbol{x}^{\text{opt}}$ such that

$$f(\boldsymbol{x}^{\text{opt}}) = \min_{x \in V} f(\boldsymbol{x})$$

where: V – set of feasible solutions

$$V = \{\boldsymbol{x}: h_i(\boldsymbol{x}) \leqslant 0; \quad i = 1, 2, 3\}$$

Design optimization was performed for a cylindrical fired heater capable of heating 215 t/h of oil from 307°C to 420°C (thermal effect 21 MW, approximately). The allowable values of the three process quantities were as follows: maximum pressure drop $q_1^{\max} = 0.05$ MPa, maximum heat flux $q_2^{\max} = 48.9$ kW/m^2, maximum thermal effect per unit volume $q_3^{\max} = 76.8$ kW/m^3.

As a reference solution, a design was defined for arbitrary assumptions. The weight of heating surface elements was calculated at 32.4 tonnes. The decision space was subsequently defined by assuming ranges of values for all decision variables. The result of the optimization was that the heating surface weight was reduced to 19.6 tonnes.

The input data and results for both calculations are shown in Table 11.1. The results supported the recommendation given in the literature [69], that in a well-designed heater, the heating surface areas in the two sections should be approximately equal.

Table 11.1

Comparison between an arbitrary and an optimal design of a fired heater

Decision variable	Assumed value	Range of values for optimization	Optimal value
λ	1.25	1.15–1.25	1.15
ψ	0.70	0.65–0.75	0.70
q_A kW/m^2	5700	5240–6400	6400
c_r m/s	1.46	1.0–1.8	1.42
c_s kg/m^{2}s	1.67	1.5–2.0	1.64
d mm	168	140, 159, 168, 194	140
e/d	1.82	1.6–2.2	2.2
Mass of piping, tons	32.4		19.6
Number of flow paths	4		6
Heating surface area, m^2			
Radiation section	467		281
Convection section	296		264

11.4 SERIES OF ROBERT-TYPE EVAPORATORS

Robert-type evaporators (see Example 9.4 in Section 9.3) are common in sugar manufacture as components of multiple-effect evaporator systems which are the central parts of sugar plants. The demand for Robert-type evaporators in the Polish sugar industry is quite large. This arises from the needs of 78 domestic sugar plants, along with the export of numerous new or modernized sugar plants abroad. The demand is differentiated qualitatively as the evaporators are used in plants of different capacities using various raw materials (beets of different processing value) and diverse processes. Even in a multiple-effect evaporator in a specific sugar plant, the process parameters vary depending on what evaporation stage is involved.

This general demand picture justifies the need to design a series of evaporators. In addition, a tendency to build sugar plants of increasing capacities should be taken into account. As the design of a series is a laborious and costly undertaking, it needs to be justified by long term use of the results. Consequently, the series should meet not only existing demand but also needs which may arise in the future. In other words, a demand forecast accounting for future sugar plant development should be made.

Another peculiarity of evaporators is that it is difficult to formulate the problem of series quality. The quality of each particular unit can, of course, be quantitatively evaluated using, for example, the evaporator weight as a criterion. No criterion, however, is available for measuring the quality of the series, with mutual relations between the properties of various sizes taken into account. Even the evaluation scheme used for the series of double-pipe heat exchangers could not be used here, as no criterion has been found which coud account for the evaporator design and the process at the same time. In other words, the evaporation process which consists of heat transfer under natural circulation of the solution processed, does not allow the designer a free choice of the values of important quantities.

Despite this difficulty, a series of evaporators was designed with the aid of the general decision-making model explained is Section 7.3, adjusted to the particular features of the problem. This made it possible to adjust the series to the differentiated demand, while also taking the quality factor into account.

The demand forecast was based on the analysis of thermal systems of sugar plants (Fig. 11.5). A thermal system can be split into four parts corresponding to the four different stages of the process:

(1) Sugar extraction from cossettes (i.e. sliced beets). Feed water supplied to the extractor and the extractor itself are continuously heated.

(2) Juice purification including the operations of liming, carbonization, filtration, etc. Between these operations, the juice is heated.

(3) Juice thickening in a multiple-effect evaporator where excess water is evaporated. The vapours are utilized in the heating of other parts of the plant.

(4) Multistage crystallization of sugar in the so-called sugar end. A considerable amount of heat is consumed here for the evaporation of the residual water and for crystal drying.

Using a relevant mathematical model and a computer program for thermal system design [4], a study was made of the effect of the following factors on the evaporation process in the evaporator station [79]:

- processing capacity of the plant (between 3000 and 9000 tonnes of beets per day),
- sugar content in cossettes (13.5% to 18%),
- juice draft in the extractor (between 110 kg and 120 kg per 100 kg beets),
- extent of scale formation at the heating surfaces,

- application or otherwise of the juice decalcification process,
- process layout at the sugar end (two cases of a three-stage process as well as the pattern of four- and five-stage processes).

The scope of this analysis encompasses domestic and export needs for a period of several years ahead.

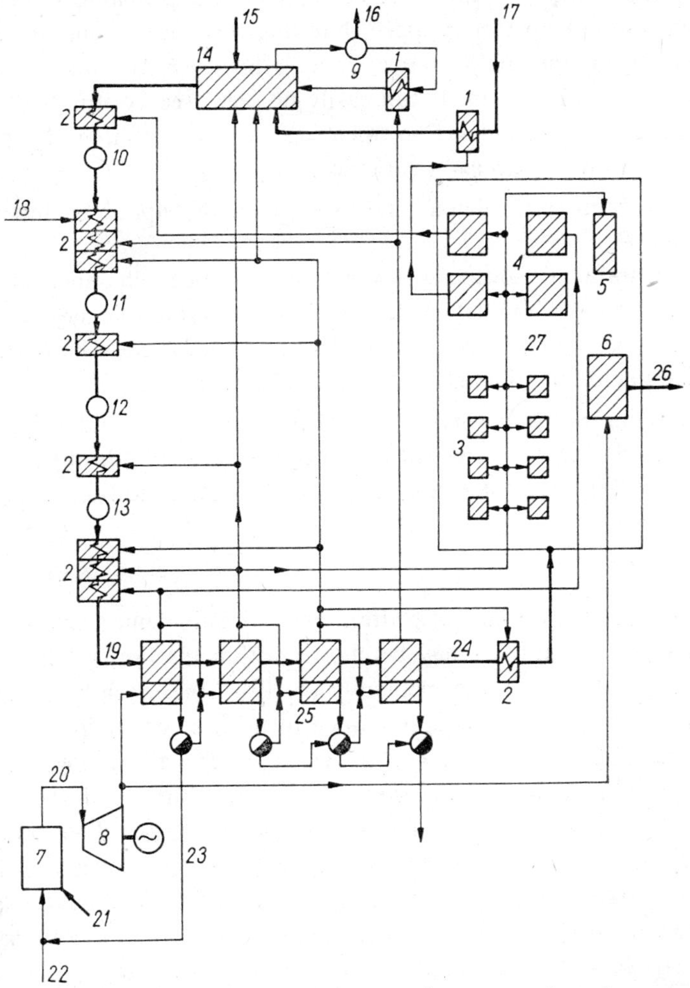

Fig. 11.5 Simplified scheme of a thermal system in a sugar plant. 1 – water heaters, 2 – juice heaters, 3 – tanks containing intermediate products, 4 – vacuum pans, 5 – melter, 6 – sugar dryer, 7 – boiler, 8 – turbine, 9 – pulp press, 10 – preliming, 11 – main liming, 12 – carbonatation, 13 – filters, 14 – extractor, 15 – cossettes, 16 – exhausted cossettes (pulp), 17 – water, 18 – condensate, 19 – thin juice, 20 – steam, 21 – fuel, 22 – feedwater, 23 – power plant, 24 – thick juice, 25 – multiple evaporator station, 26 – sugar, 27 – sugar end.

It follows from the mathematical model of the Robert-type evaporator (Example 9.4) that eight process quantities are required to define the operation mode of the unit; concentration and temperature of the solution at inlet and outlet, heating steam temperature, heat transfer coefficient, mass flow of the solution at the inlet, and heat loss coefficient. The results of the demand analysis can thus be treated as a probability measure in eight-dimensional operation space, but this form is difficult to interpret and use. Using the mathematical model, however, the analysis can be presented in another form. It can easily be seen (see Section 9.3) that the constraint system includes four relations directly defining the effect of process quantities on decision variables:

- constraint (9.12) defines the lower limit of vapour chamber diameter D_2;
- constraint (9.13) defines lower limit of vapour chamber height H_2;
- constraint (9.14) defines the smallest allowable thickness of heating chamber wall g_1 (and, as has been assumed for the sake of structural simplicity, the thickness of vapour chamber wall g_2).

The remaining decision variables, namely diameter D_1 and height H_1 of the heating chamber, are related to the operation mode through equations (9.9)–(9.11). For structural reasons, an additional condition must hold

$$D_2 \geqslant D_1 \tag{11.19}$$

It can thus be seen that the operation mode directly defines H_2, g_1 and g_2. Variables D_1, D_2 and H_1 may be chosen within the limits determined by equations (9.10), (9.11) and inequality (11.19). Consequently, the demand forecast can be expressed in a simplified form as a correlation of diameter D_2 and heating surface area F (Fig. 11.6). As the diagram given in Fig. 11.6 is hardly readable, it is reproduced, in a still further simplified form, in Fig. 11.7. Following this correlation, vapour chamber diameters should be chosen in a manner facilitating the application of evaporators at different operating modes occurring in the various sugar plants. It should be pointed out, however, that it would be uneconomical to adopt, for all the evaporators, D_2 values as required for the first evaporator effect (i.e. the first stage in the evaporation process). Namely, evaporators suited to other effects – making up more than 3/4 of the total number of evaporators produced – may have a smaller D_2 and thus smaller weight. It is thus advisable to include, in the series, evaporator designs suited to the first evaporation stage and other designs suited to the remaining stages (considered jointly).

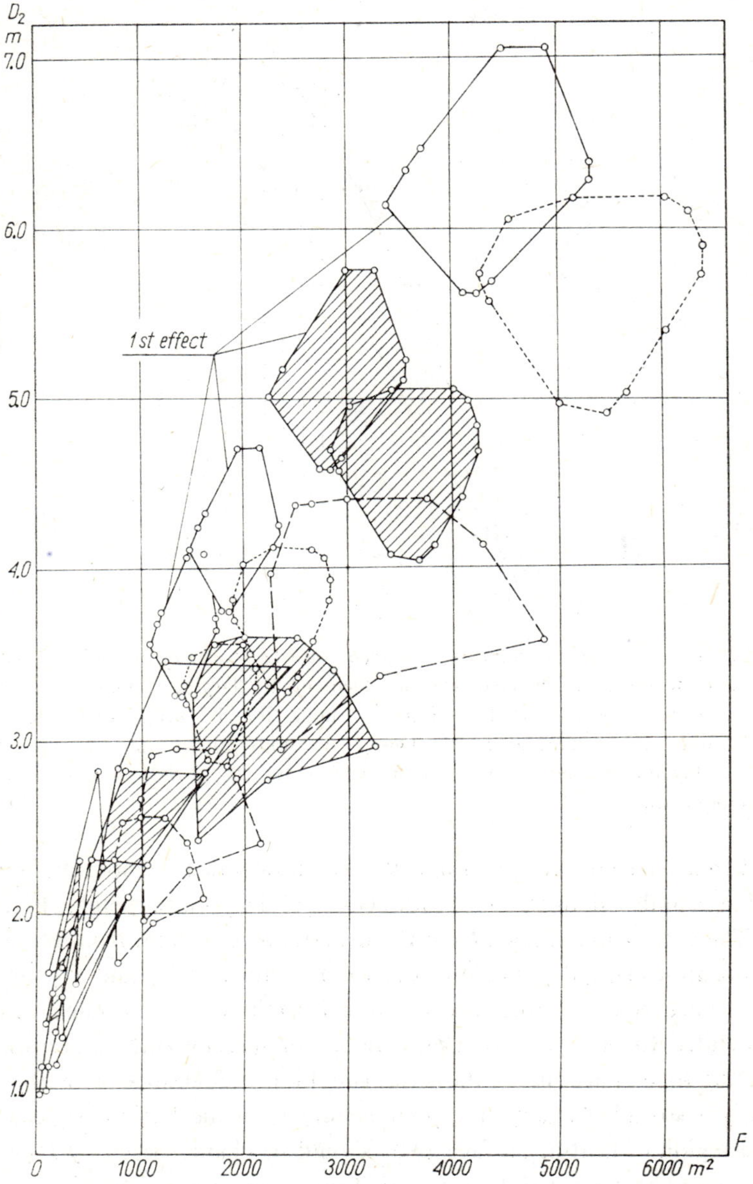

Fig. 11.6 Distribution of demand for various evaporator designs in sugar plants of 3, 4, 6 and 9 thousand tons of beet daily processing capacity. The area of each polygon corresponds to one sugar plant capacity and one evaporation stage. Shaded polygons represent demand in sugar plants of 6000 tons/day capacity.

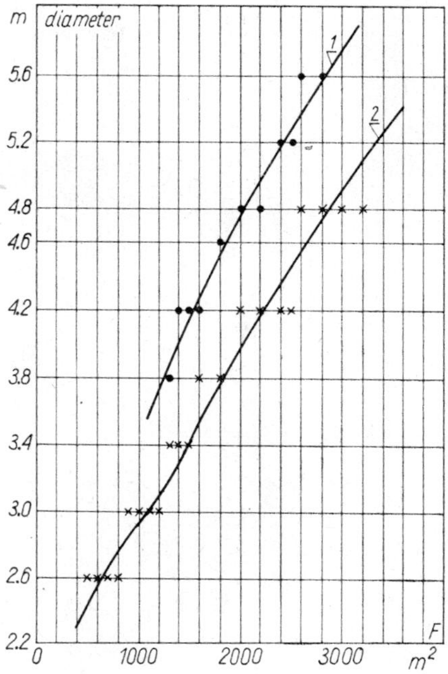

Fig. 11.7 Principle of adjusting diameters of evaporators in the series to the values identified in demand analysis. Demand for largest vapour chamber diameters according to Fig. 11.6: 1 – units for 1st evaporation stage, 2 – units for remaining stages. Design decisions concerning the series: × – heating chambers and vapour chambers ($D_1 = D_2$), • – enlarged vapour chambers.

From the solution of Example 9.4 we know how to choose the optimal heating chamber diameter D_1 (with respect to evaporator weight, Fig. 9.6).

The conclusions drawn from the demand forecast and from the design optimization make it possible to determine diameters D_1 and D_2 for arbitrary values of the heating surface area F for both the 1st effect and for units suited to other evaporation stages. For heating surface areas corresponding to 1st effect units, the series should include designs with $D_2 > D_1$ (this is a so-called enlarged vapour chamber), while basic designs in the entire region of heating surface areas should be characterized by $D_2 = D_1$.

It must be conceded, however, that no unique answer has been found to the question of how many designs and what heating surface areas should be included in the series, because no relevant evaluation criterion has been found. Instead, the decision has been made on the basis of engineering experience and intuition. Namely, the number of evaporator sizes and their differentiation have been determined so that heating surface areas form a series of easy-to-remember numbers with a 100-m^2 increment at lower values and a 200-m^2 increment at higher values [79, 82]. In this manner, the optimization of the evaporator series has been replaced by the optimization of individual evaporator designs which form a series well-suited to the prospective demand, but in which heating surface areas are in fact arbitrarily determined. In Fig. 11.7, the structure of the series is visualized against the background of values of diameter D_2 as determined in the demand forecast. A detailed presentation of the series is given in Table 11.2.

Table 11.2

Series of Robert-type evaporators for the sugar industry

No.	Heating surface/area F	Heating chamber		Vapour chamber version basic		Vapour chamber version enlarged	
	m^2	D_1 mm	H_1 mm	D_2 mm	H_2 mm	D_2 mm	H_2 mm
1	500	2600	2000	2600	5000		
2	600	2600	2500	2600	6000		
3	700	2600	2700	2600	6500		
4	800	2600	3000	2600	7000		
5	900	3000	2700	3000	5000		
6	1000	3000	2850	3000	6000		
7	1100	3000	3150	3000	6000	3800	7000
8	1200	3000	3300	3000	6500	3800	7000
9	1300	3400	2850	3400	5000		
10	1400	3400	3000	3400	6000		
11	1500	3400	3300	3400	6000	4200	7000
12	1600	3800	2850	3800	6000		
13	1800	3800	3150	3800	7000	4600	7000
14	2000	4200	2850	4200	6000		
15	2200	4200	3150	4200	6500	4800	7000
16	2400	4200	3300	4200	7000	5200	7000
17	2500	4200	3500	4200	7000	5200	7000
18	2600	4800	3000	4800	6500	5600	7000
19	2800	4800	3000	4800	6500	5600	7000
20	3000	4800	3300	4800	7000	5800	7000
21	3200	4800	3500	4800	7000	6000	7000

APPENDIX 1

Selected topics from set theory

In this Appendix, the most important concepts, definitions and symbols from set theory are presented to the extent that they have been used in this book. Readers interested in further study of the subject are referred to the standard texts [17, 41, 59].

Set algebra

The idea of a *set* can be regarded as one of the fundamental, non-definable concepts of mathematics. Set theory is a branch of mathematics concerned with the properties of sets, independent of the nature of the objects which form the sets.

The objects belonging to a set are referred to as *set elements.* A statement specifying that a is an element of set A can be written as

$$a \in A$$

and if a is not an element of set A

$$a \notin A$$

A set consisting of elements $a_1, a_2, \ldots, a_n$ can be written as

$$\{a_1, a_2, \ldots, a_n\}$$

A set including no elements is called a null-set, or an *empty set.* If every element of set A is also an element of set B, then set A is said to be *included in B*, or it is a *subset of B*

$$A \subset B$$

The *union of two sets A* and B is a set of all elements of set A and all elements of set B containing no other elements

$$A \cup B$$

The *intersection of sets A* and B is a set of elements belonging to set A and to set B simultaneously

$$A \cap B$$

In a particular application, the reasoning based on the concepts of set theory is usually concerned with elements and subsets of a definite set which is called a *space*. For instance, arithmetic is concerned with the space of real numbers, and Euclidean geometry is concerned with points in Euclidean space.

In a given space X, the expression $\phi(x)$ in which variable x appears and which becomes a true statement or a false statement when an arbitrary element of space X is substituted for x, is called a *statement function* of variable x. An element $y \in X$ satisfies the statement function $\phi(x)$, if statement $\phi(y)$ is true. A set of elements in space X satisfying this statement function is denoted as

$$\{x \in X:\ \phi(x)\}$$

For example, if X is a space of positive integers, then

$$\{x \in X:\ x < 2\}$$

is identical with $\{0, 1\}$.

Functions

If X and Y are arbitrary non-empty sets and to every $x \in X$ exactly one $y \in Y$ is assigned, then one can talk of

– mapping of set X onto set Y;

– a function mapping X onto Y.

The two expressions have exactly the same meaning. The statement "function f maps set X onto set Y" can be written down as

$$f\colon X \to Y$$

An element of set Y assigned by mapping f to element x of set X is called *function value*, or *image of x at mapping f*. It is denoted by $f(x)$. Function f is said to be *defined on set X*; alternatively, it is said that X is a *domain of function f*. Set Y, that is the set of values of f, is also called a *codomain of f*. It can be denoted $f(X)$.

Note: if $f\colon X \to Y$, but $f(X)$ is not identical with Y, then it is said that function f is a *mapping* of set X *onto* set Y. If $f(X) = Y$, then it is a *one-to-one mapping* of X onto Y.

For a given $f\colon X \to Y$, it may be necessary to investigate the function on a certain subset $A \subset X$. It can then be regarded as a new function $g\colon A \to Y$, with the property

$$g(x) = f(x) \qquad \text{for} \qquad x \in A$$

Function g is called *function f restricted* to set A, and may be denoted $f|A$.

If $g\colon A \to Y$ is a function defined on subset $A \subset X$ and it is known of function $f: X \to Y$ that $f|A = g$, then f is called the *expansion of function* g on set X.

Function $f: X \to Y$ is called

– a *real-valued function*, if Y is a set of real numbers (i.e. if function values are real numbers),

– a *series*, if X is a set of positive integers (i.e. $f(n)$ appears as nth term of a series).

Cartesian products

If two arbitrary objects a, b are given, then an *ordered pair* with antecedent a and consequent b can be defined and is denoted (a, b).

A *Cartesian product of sets* A and B is a set of all ordered pairs (a, b) such that $a \in A$, $b \in B$. The Cartesian product is denoted $A \times B$.

If a Cartesian product $X \times Y$ of non-empty sets X and Y is given, then expression $\phi(x, y)$ which includes variables x, y and which becomes a statement when the name of an arbitrary element of X is substituted for x while the name of an arbitrary element of Y is substituted for Y – is called a *statement function* of two variables x and y. A statement function of two variables x and y taking values from sets X and Y, respectively, can be regarded as a statement function of one variable $z = (x, y)$ taking values from Cartesian product $X \times Y$. A set of elements of product $X \times Y$ satisfying statement function $\phi(x, y)$ is denoted

$$\{(x, y) \in X \times Y: \phi(x, y)\}$$

By analogy to the Cartesian product of two sets, generalized Cartesian products of many sets can be defined. For example, a three-dimensional Euclidean space is a Cartesian product $R \times R \times R$ (or R^3) where R is a set of all real numbers. Generally, an m-dimensional Euclidean space is a Cartesian product R^m. In a Cartesian product of m sets, statement functions of m variables can be defined.

From the point of view of set theory, functions of many variables are special cases of a function of one variable. If X is an arbitrary non-empty set, then function $f: X^n \to Y$ is called a function of n variables taking values from set X; function values belong to set Y. Arguments of a function in n variables are elements of set X^n, that is a series of n terms $(x_1, x_2, \ldots, x_n)$, each term belonging to X. The value of f for argument $(x_1, x_2, \ldots, x_n)$ is denoted by $f(x_1, x_2, \ldots, x_n)$.

Finite covering

A set of elements which are also sets, is called a *set family*. If X is an arbitrary space, $\mathscr{W}$ is a family of all subsets in this space and T is an arbitrary non-empty set, then function $f\colon T \to \mathscr{W}$, assigning sets from family W to elements of set T, is called an *indexed set family*. The notation $f(t) = A_t$ means that to every $t \in T$, one set $A_t \in \mathscr{W}$ is assigned. Family $\mathscr{W}$ can thus be treated as set $\{A_t\}$, for all possible t.

A *covering of space* X is such a set family $\{\mathscr{G}_t\}$, t taking all values from a set T, that

$$\bigcup_{t \in T} \mathscr{G}_t = X$$

Set T may be infinite. The existence of a covering as defined above implies that each element of space X belongs to at least one set from family $\{\mathscr{G}_t\}$.

A *finite covering of space* X is such a finite set family $\{\mathscr{G}_t;\ t = 1, 2, \ldots \ldots n\}$ that

$$\bigcup_{t=1}^{n} \mathscr{G}_t = X$$

APPENDIX 2

Selected topics from vector and matrix calculus

In this Appendix, the most important concepts, definitions and symbols of vector and matrix calculus are presented to the extent they are needed in numerical methods of optimization. For an extension of this treatment and for specific practical examples, the reader is referred to the literature [5, 67].

Vectors

The subject of matrix calculus is that of operations on linear forms, for example on systems of linear equations. The advantage of using vectors and matrices is the notation which allows convenient manipulation of multi-variable functions.

The two sets of numbers $x_1, x_2, \ldots, x_n$ and $y_1, y_2, \ldots, y_m$ of n and m elements, respectively, can be ordered and subsequently treated as column vectors

$$\boldsymbol{x} = \begin{bmatrix} x_1 \\ x_2 \\ \vdots \\ x_n \end{bmatrix} \qquad \boldsymbol{y} = \begin{bmatrix} y_1 \\ y_2 \\ \vdots \\ y_m \end{bmatrix}$$

Vector $\boldsymbol{x}$ is regarded as n-dimensional and vector $\boldsymbol{y}$ as m-dimensional. Numbers $x_1, x_2, \ldots, x_n$ and $y_1, y_2, \ldots, y_m$ are called *components of vectors* $\boldsymbol{x}$ and $\boldsymbol{y}$. When the components are ordered in a row, a row vector is obtained. An operation transforming a column vector into a row vector, or conversely, is called transposition. The result of the transposition of vector $\boldsymbol{x}$ is

$$\boldsymbol{x}^T = [x_1, x_2, \ldots, x_n]$$

Matrices

Let components of vectors $\boldsymbol{x}$ and $\boldsymbol{y}$ be interconnected by linear relations

$$
\begin{aligned}
a_{11}x_1 + a_{12}x_2 + \ldots + a_{1n}x_n &= y_1 \\
a_{21}x_1 + a_{22}x_2 + \ldots + a_{2n}x_n &= y_2 \\
\vdots \qquad\qquad\qquad\qquad & \quad \vdots \\
a_{m1}x_1 + a_{m2}x_2 + \ldots + a_{mn}x_n &= y_m
\end{aligned}
$$

Such an interconnection is called a *system of linear equations.* The system is *overdefined* if $m > n$, and it is *indefinite* if $m < n$. Coefficients (or entries) a_{ij} can be symbolized by a matrix

$$
\mathbf{A} = \begin{bmatrix} a_{11} a_{12} & \ldots & a_{1n} \\ a_{21} a_{22} & \ldots & a_{2n} \\ \vdots & & \vdots \\ a_{m1} a_{m2} & \ldots & a_{mn} \end{bmatrix} = [a_{ij}]
$$

and the system of linear equations can be expressed as

$$
\mathbf{A}\boldsymbol{x} = \boldsymbol{y}
$$

Evidently, this notation includes certain rules of matrix and vector operations which will be explained later.

A *transposed matrix* $\mathbf{A}^T$ is a matrix obtained from matrix $\mathbf{A}$ by interchanging its rows and columns

$$
\mathbf{A}^T = \begin{bmatrix} a_{11} & a_{21} & \ldots & a_{m1} \\ a_{12} & a_{22} & \ldots & a_{m2} \\ \vdots & & & \vdots \\ a_{1n} & a_{2n} & \ldots & a_{mn} \end{bmatrix}
$$

A *quadratic,* or *square, matrix* is a matrix of equal number of rows and columns ($m = n$).

A *diagonal matrix* $\mathbf{D}$ is a quadratic matrix, the entries of which in the major diagonal are different from zero, while the remaining entries are zeros

$$
\mathbf{D} = \begin{bmatrix} a_{11} & 0 & \ldots & 0 \\ 0 & a_{22} & \ldots & 0 \\ \vdots & & & \vdots \\ 0 & 0 & \ldots & a_{nn} \end{bmatrix}
$$

When all the entries in the diagonal of the diagonal matrix are equal to 1, the matrix is called a *unit matrix*

$$\mathbf{I} = \begin{bmatrix} 1 & 0 & \dots & 0 \\ 0 & 1 & \dots & 0 \\ \vdots & & & \vdots \\ 0 & 0 & \dots & 1 \end{bmatrix}$$

A matrix with the entries in the major diagonal and above (or below) it equal to zero is called a *triangular matrix*. The two possible variants of this matrix are: upper triangular matrix

$$\mathbf{T}_1 = \begin{bmatrix} 0 & a_{12} & \dots & a_{1n} \\ 0 & 0 & \dots & a_{2n} \\ \vdots & & & \vdots \\ 0 & 0 & \dots & 0 \end{bmatrix}$$

and lower triangular matrix

$$\mathbf{T}_2 = \begin{bmatrix} 0 & 0 & \dots & 0 \\ a_{21} & 0 & \dots & 0 \\ \vdots & & & \vdots \\ a_{n1} & a_{n2} & \dots & 0 \end{bmatrix}$$

An *inverse matrix* is a concept related to solving systems of linear equations. If we determine unknown variables x_i from the system of linear equations presented above, the solution can generally be expressed as

$$x_1 = b_{11}y_1 + b_{12}y_2 + \dots + b_{1m}y_m$$
$$\vdots$$
$$x_n = b_{n1}y_1 + b_{n2}y_2 + \dots + b_{nm}y_m$$

or in the vector – matrix form

$$\boldsymbol{x} = \mathbf{A}^{-1}\boldsymbol{y}$$

Matrix $\mathbf{A}^{-1} = [b_{ij}]$ is called the *inverse of matrix* **A**.

It should be pointed out that the above form of a solution of a system of linear equations is not universal, that is, an inverse matrix does not always exist. A necessary condition for the inverse matrix $\mathbf{A}^{-1}$ to exist is that **A** is a non-singular matrix (a singular matrix is a matrix the determinant of which is equal to zero).

Vector and matrix operations

An inner product of two vectors is well-defined only if both vectors are of the same dimension. For given vectors

$$\boldsymbol{x}_1 = \begin{bmatrix} x_{11} \\ x_{21} \\ \vdots \\ x_{n1} \end{bmatrix} \qquad \boldsymbol{x}_2 = \begin{bmatrix} x_{12} \\ x_{22} \\ \vdots \\ x_{n2} \end{bmatrix}$$

their inner product is a number defined as

$$z = x_{11}x_{12}+x_{21}x_{22}+ \ldots +x_{n1}x_{n2} = \sum_{i=1}^{n} x_{i1}x_{i2}$$

In vector form, the inner product is denoted

$$z = \boldsymbol{x}_1^T\boldsymbol{x}_2 = \boldsymbol{x}_2^T\boldsymbol{x}_1$$

Addition (or subtraction) of matrices **A** and **B** is defined for matrices of equal dimensions only. The entries of the resulting matrix **C** are sums or differences of the entries of matrices **A** and **B**

$$\mathbf{C} = \mathbf{A}\pm\mathbf{B} = [c_{ij}]$$

$$c_{ij} = a_{ij}\pm b_{ij}$$

Multiplication of a matrix by a scalar is equivalent to the multiplication of all entries by the same scalar

$$k\mathbf{A} = \mathbf{A}k = [ka_{ij}]$$

As every vector can be treated as a one-column (or a one-row) matrix, the same is also valid for vectors.

A product of matrices **A** and **B** is well-defined only if the number of rows in matrix **B** is equal to the number of columns in matrix **A**. The entries of the resulting matrix **C** are inner products of rows of matrix **A** by columns of matrix **B**

$$\mathbf{C} = \mathbf{AB} = [c_{ij}]$$

where: **A** – $m\times n$ matrix, **B** – $n\times p$ matrix, and

$$c_{ij} = \sum_{r=1}^{n} a_{ir}b_{rj} \qquad i = 1, \ldots m; \qquad j = 1, \ldots, p$$

Note that, in general, **AB** does not equal **BA**.

Owing to the properties of matrix multiplication, equation $\mathbf{AB} = \mathbf{AC}$ does not necessarily mean that $\mathbf{B} = \mathbf{C}$.

The result of the multiplication of a matrix by its inverse is a unit matrix

$$\mathbf{AA}^{-1} = \mathbf{I}$$

The result of the transposition of a product of two matrices is identical to the product of the transposed matrices, multiplied in a reverse order

$$(\mathbf{AB})^T = \mathbf{B}^T\mathbf{A}^T$$

APPENDIX 3

Selected numerical methods

In this Appendix, basic numerical methods are comprehensively reviewed, namely interpolation, approximation, solving of nonlinear equations with one unknown and solving of systems of equations with many unknowns. Familiarity with these methods is necessary for an understanding of direct methods of finding the extremum of a function. This is because some of the extremum-seeking methods are explicitly based on interpolation or approximation and the iterative solving of equations or of systems of equations. Numerical methods are also of primary importance in preparing mathematical models for computerized calculations. They are, for instance, very helpful in transforming the relations given in a graph or table form into an easy to program form.

The review presented below is meant only to aid the reader in understanding the contents of this book and should not be regarded as an exhaustive treatment of the subject. For a systematic treatment of numerical methods, the pertinent literature should be consulted [18, 64].

A.3.1 Interpolation

If function $b(\boldsymbol{x})$ is defined by a set of values assigned to a set of points $\boldsymbol{a}_1, \boldsymbol{a}_2, \ldots, \boldsymbol{a}_r$ in space X (grid points), then the interpolation problem consists of determining function values outside the given points, with an estimate of the error involved. The solution is understood as certain function $y(\boldsymbol{x})$ which at grid points takes the same values as $b(\boldsymbol{x})$.

The example of a one-dimensional problem, i.e. interpolation of a function of one variable, is discussed below. All the formulae applicable to functions of many variables can be regarded as extensions of this fundamental case.

Assume that the unknown function $y(x)$ should be constructed on the basis of known values $b(a_1), b(a_2), \ldots, b(a_r)$. Its general form is

$$y(x) = \sum_{j=1}^{r} l_j(x)b(a_j) \tag{A.1}$$

where: $l_j(x), j = 1, 2, \ldots, r$ are functions which can be chosen in accordance with particular interpolation methods. These functions are usually required to be polynomials, so that $y(x)$ is also a polynomial.

In the case of the Lagrange interpolation formula, it is assumed that $l_j(x)$ contains a factor

$$(x-a_1)(x-a_2) \ldots (x-a_{j-1})(x-a_{j+1}) \ldots (x-a_r) \tag{A.2}$$

This factor ensures that $l_j(a_k) = 0$ for every $j \neq k$. If it could be simultaneously arranged that $l_j(a_j) = 1$, then $y(x)$ would exactly reproduce $b(x)$ at all grid points. This can be achieved by defining

$$l_j(x) = \frac{(x-a_1)(x-a_2) \ldots (x-a_{j-1})(x-a_{j+1}) \ldots (x-a_r)}{(a_j-a_1)(a_j-a_2) \ldots (a_j-a_{j-1})(a_j-a_{j+1}) \ldots (a_j-a_r)}$$

The above expression is known as the *Lagrange interpolation polynomial.* The complete Lagrange interpolation formula is thus

$$y(x) = \sum_{j=1}^{n} \frac{(x-a_1) \ldots (x-a_{j-1})(x-a_{j+1}) \ldots (x-a_r)}{(a_j-a_1) \ldots (a_j-a_{j-1})(a_j-a_{j+1}) \ldots (a_j-a_r)} \, b(a_j)$$

The error of Lagrange interpolation can be estimated from the formula

$$\delta(x) = \frac{p_r(a)}{n^r!} \, b^{(r)}(\xi)$$

where: $p_r(x) = \prod_{i=1}^{n} (x-a_i)$; $b^{(r)}$ – rth derivative of function b, ξ – unknown function of x, taking values in the range $[a_1, a_r]$.

A variety of interpolation methods can be found in the literature. Some, like the Lagrange method, making it possible to determine the analytical form of the interpolating function. Others are based on finite difference methods, reducing the interpolation to performing arithmetical operations on the values of variables and functions at grid points. Various interpolation subroutines can be found in mathematical program libraries, such as SSP (IBM) or MATHLIB (CDC).

Example A.1

Determine a Lagrange interpolation polynomial for function $b(x)$ whose values are given in Table A.1.

Polynomials $l_i(x)$, $i = 1, 3, 4$ can be determined from formula (A.2). As $b(a_2) = 0$, $l_2(x)$ may be ignored.

Table A.1

Function values for the interpolation example

i	1	2	3	4
a_i	0	0.1	0.3	0.5
$b(a_i)$	−0.5	0	0.2	1

$$l_1(x) = \frac{(x-0.1)(x-0.3)(x-0.5)}{(-0.1)(-0.3)(-0.5)} = \frac{x^3-0.9x^2+0.23x-0.015}{-0.015}$$

$$l_3(x) = \frac{x(x-0.1)(x-0.5)}{0.3\times 0.2(-0.2)} = \frac{x^3-0.6x^2+0.05x}{-0.012}$$

$$l_4(x) = \frac{x(x-0.1)(x-0.3)}{0.5\times 0.4\times 0.2} = \frac{x^3-0.4x^2+0.03x}{0.04}$$

Consequently, the required Lagrange polynomial assumes the form

$$y(x) = l_1(x)b(a_1)+l_2(x)b(a_2)+l_3(x)b(a_3)+l_4(x)b(a_4) =$$

$$= \frac{125}{3}x^3-30x^2+\frac{73}{12}x-0.5$$

A.3.2 Approximation

The approximation problem is analogous to that formulated in Section A.3.1. It is not, however, required that functions b and y coincide at all grid points. The approximation of functions of one variable is discussed below.

If the approximating function y assumes form (A.1), then owing to the relaxed requirements, polynomials $l_j(x)$ of a degree lower than $r-1$ can be accepted. This is particularly important when the given values of b are associated with random errors (for example, measurement errors), Fig. A.1. Approximation by polynomials of a sufficiently low degree makes it possible to smooth the function, while applying interpolation implies that the function values supplied are exact when they could, in fact, be inaccurately determined. If the data are associated with errors, then the approximation makes it possible to better reproduce the derivative of function b than the interpolation.

One possible form of the required approximating function is

$$y_m(x) = \sum_{j=0}^{m} c_j(m)\phi_j(x)$$

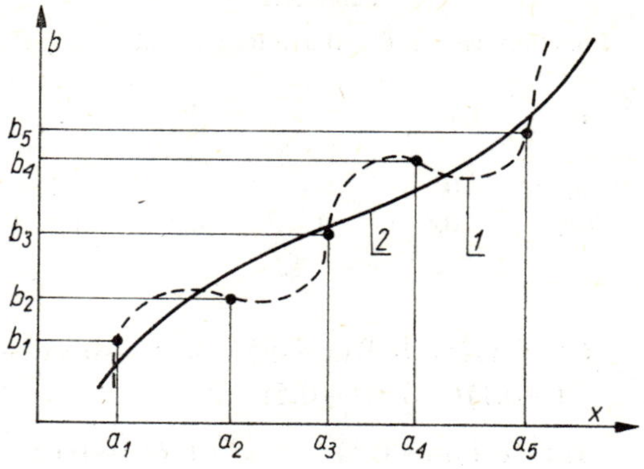

Fig. A.1 Qualitative differences between interpolation and approximation in the case of a function defined by inacurate values. 1 – interpolation polynomial, 2 – approximating function $y(x)$.

where: $\phi_j(x)$ – functions analogous to $l_j(x)$ in the interpolation formula, $c_j^{(m)}$ – constants, m – degree of approximation.

Function values associated with errors at points $x_1, x_2, \ldots, x_r$ will hereafter be denoted $b_1, b_2, \ldots, b_r$.

The approximation problem consists of finding such function y that $y(x_i) \approx b_i$, for $i = 1, 2, \ldots, r$. To make this formulation unique, additional requirements are needed. Most frequently, recourse is made to a least-squares approximation. This relies on the requirement that the following expression, which is a measure of error of reproducing $b(x)$, attains its minimum:

$$\Phi\left(c_0^{(m)}, c_1^{(m)}, \ldots, c_m^{(m)}\right) = \sum_{i=1}^{r} w(x_i)\,[b_i - y(x_i)]^2$$

$$= \sum_{i=1}^{r} w(x_i)\left[b_i - \sum_{j=0}^{m} c_j^{(m)}\phi_j(x_i)\right]^2 \qquad \text{(A.3)}$$

The determination of the approximating function $y_m(x)$ consists of finding the values of coefficients $c_j^{(m)}$, $j = 0, 1, \ldots, m$, corresponding to the minimum of the function Φ. Function w in formula (A.3) is a so-called weighing function which takes on non-negative values reflecting the importance of errors at individual grid points. It is frequently assumed that for every x, $w(x) = 1$, which is equivalent to attaching equal importance to all grid points.

It can be easily seen that if $r = m+1$, then error measure Φ can be reduced to zero, for example by assuming that $y_m(x)$ is a Lagrange interpolation polynomial. The smoothing effect, however, can be attained only if $r > m+1$. Functions $\phi_j(x)$ can be polynomials, but also – depending on estimated properties of $b(x)$ or desirable properties of $y(x)$ – exponential, trigonometric or other functions.

Methods of finding coefficients $c_j^{(m)}$ are presented in numerous textbooks. Most mathematical program packages include programs for finding either coefficients $c_j^{(m)}$ in the polynomial representation, or coefficients in the resulting polynomial

$$\sum_{j=0}^{m} c_j^{(m)} \phi_j(x)$$

The degree of approximation m is usually chosen so as to ensure that maximum error $|b_i - y(x_i)|$ does not exceed a certain allowable value. Generally, the question of whether such a condition can be satisfied is resolved by examination of the results.

Example A.2

Function $b(x)$ is defined by its values approximately measured at five points, Table A.2. Determine a second-degree polynomial approximation of that function:

$$y(x) = a_0 + a_1 x + a_2 x^2$$

Table A.2

Function values for the approximation example

i	1	2	3	4	5
x_i	0	1	2	3	4
b_i	1.000	3.850	6.500	9.350	12.050

As only three coefficients should be found and the number of grid points is small, we can rely on the so-called *system of normal equations.* It defines the conditions of existence of a minimum of function Φ according to formula (A.3)

$$\frac{\partial \Phi}{\partial c_j^{(m)}} = 0, \qquad j = 0, 1, \ldots, m$$

Assuming $w(x) = 1$ we obtain

$$\Phi(a_0, a_1, a_2) = \sum_{i=1}^{5} [b_i - (a_0 + a_1 x_i + a_2 x_i^2)]^2$$

Setting the derivative of Φ, with respect to each a_i, to zero produces the system of normal equations

$$-2\sum_{i=1}^{5}[b_i-(a_0+a_1x_i+a_2x_i^2)]=0$$

$$-2\sum_{i=1}^{5}\{[b_i-(a_0+a_1x_i+a_2x_i^2)]x_i\}=0$$

$$-2\sum_{i=1}^{5}\{[b_i-(a_0+a_1x_i+a_2x_i^2)]x_i^2\}=0$$

Upon substituting data from Table A.2 and making some simplifications we obtain

$$\begin{aligned}5a_0+10a_1+30a_2 &= 32.75\\ 10a_0+30a_1+100a_2 &= 93.10\\ 30a_0+100a_1+354a_2 &= 307.0\end{aligned}$$

The solution to this system of equations is

$$a_0 = 1.046, \qquad a_1 = 2.728, \qquad a_2 = 0.008$$

and the required approximating function

$$y(x) = 1.046+2.728x+0.008x^2$$

As can be seen from Table A.3, the values of this function at the grid points are sufficiently close to the given data.

Table A.3

Approximation results

i	1	2	3	4	5
$y(x_i)$	1.046	3.782	6.534	9.311	12.086
b_i	1.000	3.850	6.500	9.350	12.050

A.3.3 Solution of nonlinear equations

If $b(x)$ is a nonlinear function, then the solution to equation $b(x) = 0$ can be expressed by a finite formula in exceptional cases only, e.g. when b is a second- or third-degree polynomial. Most frequently, the equation can be solved only approximately, with the aid of some *iterative method.*

The essential idea of an iterative method is to compute increasingly accurate consecutive approximations to the solution, until the equation is satisfied (at a certain level of accuracy). It should be added that the process starts from a certain initial approximation. The general form of the iteration formula is

$$x_{i+1} = F_i(x_i, x_{i-1}, \dots, x_{i-k+1})$$

This is a so-called *k-point formula* in which the calculation of the $(i+1)$th approximation is based on the knowledge of k preceding approximations. The knowledge of an approximation should be understood here to mean not only knowledge of x and corresponding $b(x)$ values, but in some methods also of the values of the derivatives of b. If the form of F_i is independent of i, then the iteration formula is regarded as a stable one; otherwise, it is called an unstable iteration formula.

The construction of an iteration formula can be based on various principles. For example, the Lagrange interpolation formula for an inverse function to $b(x)$ can be used (such a formula can be constructed for an arbitrary number of points greater than 1). Two methods are in frequent use in practice, the *secant method* using a two-point formula and the *Newton–Raphson method* which employs a one-point formula.

The secant method can be used when two initial approximations x_1 and x_2 are available such that $b(x_1)\,b(x_2) < 0$, that is function values at those points must be of opposite signs. Points x_1 and x_2 are subsequently used in constructing a linear interpolation formula for $b(x)$, from which a new estimate of x is obtained, Fig. A.2:

$$x_3 = \frac{y_2}{y_2 - y_1}\,x_1 + \frac{y_1}{y_1 - y_2}\,x_2$$

In the next iterative step, x_3 can be adopted as a new known point and either x_1 or x_2 is eliminated depending on which function value is of the

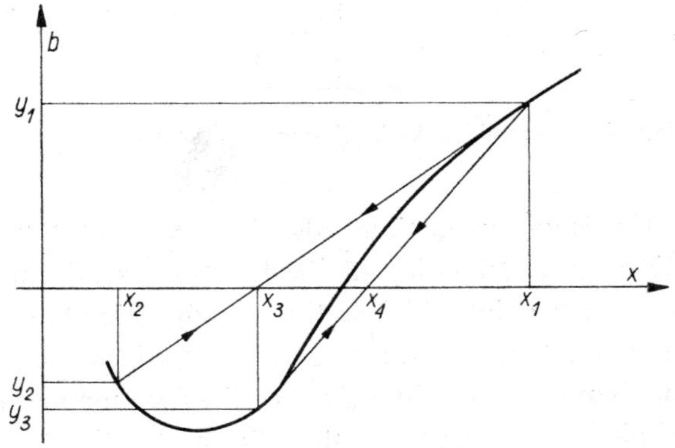

Fig. A.2 Principle of equation solving by secant method.

same sign as $b(x_3)$. In the case visualized in Fig. A.2, x_2 should be eliminated and thus

$$x_4 = \frac{y_3}{y_3 - y_1} x_1 + \frac{y_1}{y_1 - y_3} x_3$$

The procedure presented above is an example of an unstable iterative formula. Multiple repetition of calculations makes it possible to solve equation $b(x) = 0$ at an arbitrary accuracy level, as the error of a linear interpolation formula can be easily estimated. In Fig. A.3, a simplified logic diagram of the secant method is given.

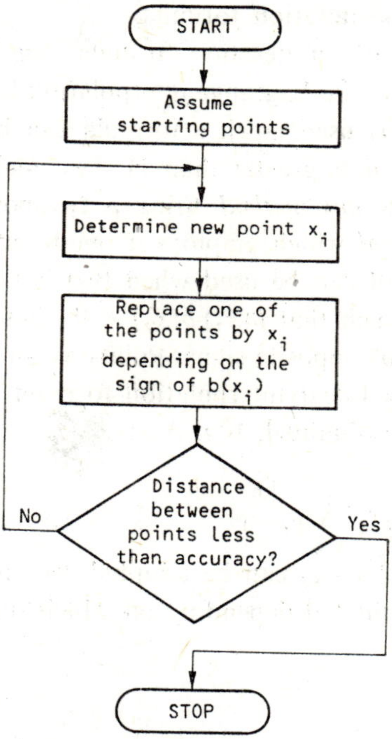

Fig. A.3 Simplified logic diagram of the secant method.

The Newton–Raphson method, also known as the tangent method, is based on linear approximation of the function exactly in the direction of its derivative, at a point which is the current estimate of the solution, Fig. A.4. The new estimate is the value of x at which the approximating expression equals zero. Once the values of the function and its first derivative at this point have been calculated, the next estimate is found, and so forth. The iteration formula is thus a stable one:

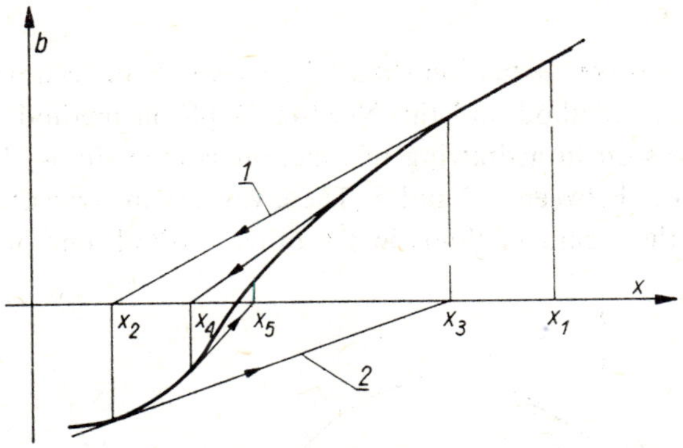

Fig. A.4 Principle of equation solving by Newton–Raphson method. 1, 2 – tangent lines at points x_1 and x_2.

$$x_{i+1} = x_i - \frac{b(x_i)}{b'(x_i)}$$

Similarly as in the secant method, the accuracy of the solution can be easily estimated. Certain properties of the function appearing in the equation $b(x) = 0$ may cause the Newton–Raphson method to fail, Fig. A.5. It can be said of this method that it might be non-convergent, as opposed to the secant method which is always convergent.

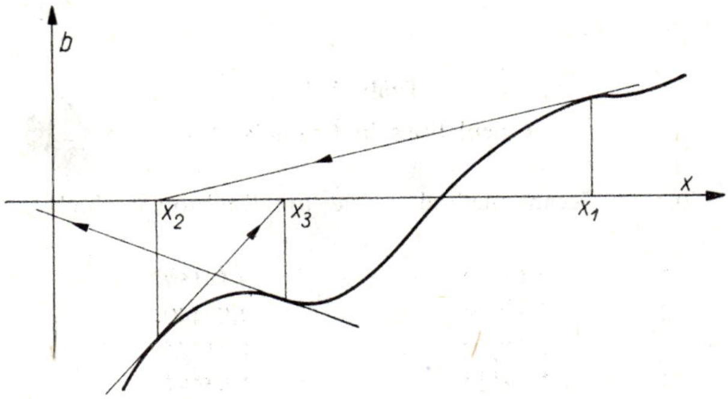

Fig. A.5 Example of a non-convergent iterative process by Newton–Raphson method.

It is worth noting that the numerical properties of the iteration formulae of the two methods are different. When applying the secant method, the solution is approached more slowly then by the Newton–Raphson method. The latter method, however, requires that the values of the function and its derivative are calculated, while in the secant method, no calculation of the derivative is needed.

Example A.3

Find a positive root of equation $\sin x - \frac{1}{2}x = 0$ with an accuracy of 1×10^{-5} using the secant method and the Newton–Raphson method.

It can be seen in a drawing of function $y(x) = \sin x - \frac{1}{2}x$ (Fig. A.6), that the root lies between $\pi/2$ and π. These values can serve as initial approximations in the secant method; in the other method, one of them can be

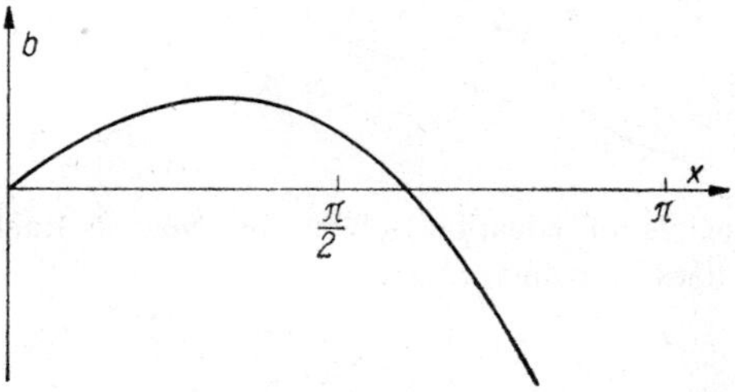

Fig. A.6 Function $b(x) = \sin x - \frac{1}{2}x$.

used as a starting point. Note that when using the Newton–Raphson method, it is necessary to calculate the values of the derivative $y'(x) = \cos x - \frac{1}{2}$. Using the iterative formulae of both methods as presented earlier we obtain consecutive estimates of the solution shown in Table A.4.

Table A.4

Calculations in Example A.3

No.	Secant method	Newton–Raphson method
1	3.14159	3.14159
2	1.57079	2.09440
3	1.75960	1.91322
4	1.84420	1.89567
5	1.87701	1.89549
6	1.88895	1.89549
7	1.89320	
8	1.89469	
9	1.89521	
10	1.89540	
11	1.89546	
12	1.89548	
13	1.89549	
14	1.89549	

A.3.4 Solution of systems of equations

Let a system of linear equations with many unknowns be of the form

$$\begin{cases} a_{11}x_1 + a_{12}x_2 + \ldots + a_{1n}x_n = b_1 \\ a_{21}x_1 + a_{22}x_2 + \ldots + a_{2n}x_n = b_2 \\ \vdots \qquad\qquad\qquad\qquad\qquad \vdots \\ a_{n1}x_1 + a_{n1}x_1 + \ldots + a_{nn}x_n = b_n \end{cases}$$

Using the rules of matrix algebra, we can rewrite the system as

$$\begin{bmatrix} a_{11} & a_{12} & \ldots & a_{1n} \\ a_{21} & a_{11} & \ldots & a_{2n} \\ \vdots & & & \vdots \\ a_{n1} & a_{n2} & \ldots & a_{nn} \end{bmatrix} \begin{bmatrix} x_1 \\ x_2 \\ \vdots \\ x_n \end{bmatrix} = \begin{bmatrix} b_1 \\ b_2 \\ \vdots \\ b_n \end{bmatrix}$$

A convenient symbolic notation is

$$\mathbf{A}\boldsymbol{x} = \boldsymbol{b}$$

where: $\mathbf{A} = [a_{ij}]$ – an $n \times n$ matrix containing the coefficients of the equations, $\boldsymbol{x}$ – column vector of unknowns, $\boldsymbol{b}$ – column vector of free terms or right hand sides.

If **A** is a non-singular matrix, then the solution of a system of linear equations is

$$\boldsymbol{x} = \mathbf{A}^{-1}\boldsymbol{b} \tag{A.4}$$

where: $\mathbf{A}^{-1}$ – inverse matrix of **A**.

As matrix inversion is a time-consuming operation and frequently associated with considerable numerical error (one can talk then of ill-conditioned matrix **A**), formula (A.4) is seldom used. Instead, a variety of numerical methods can be applied to such systems. As an example, the *Gauss–Seidel method* is discussed later. It has not been chosen for its particular suitability, but rather for the fact that it can also be of help in solving certain systems of nonlinear equations.

As in the case of nonlinear equations, the iterative methods used in solving systems of linear equations consist of successively improving solution estimates in the consecutive steps in which an iteration formula is used. Using matrix notation, we can describe the general form of a k-point iteration formula as

$$\boldsymbol{x}_{i+1} = \mathbf{F}_i(\boldsymbol{x}_{i-1}, \ldots, \boldsymbol{x}_{i-k})$$

where: $\mathbf{F}_i$ – certain matrix function.

In the Gauss–Seidel method, $\mathbf{F}_i$ remains unchanged with varying i, that is the iteration formula is stable; it is also of one-point type. The method is based on the observation that the coefficient matrix **A** can be regarded as a sum of a diagonal matrix and lower and upper triangular matrices:

$$\begin{bmatrix} a_{11} & 0 & 0 & \dots & 0 \\ 0 & a_{22} & 0 & \dots & 0 \\ 0 & 0 & a_{33} & \dots & 0 \\ \vdots & & & & \vdots \\ 0 & 0 & 0 & \dots & a_{nn} \end{bmatrix} + \begin{bmatrix} 0 & 0 & 0 & \dots & 0 \\ a_{21} & 0 & 0 & \dots & 0 \\ a_{31} & a_{32} & 0 & \dots & 0 \\ \vdots & & & & \vdots \\ a_{n1} & a_{n2} & a_{n3} & \dots & 0 \end{bmatrix} +$$

$$+ \begin{bmatrix} 0 & a_{12} & a_{13} & \dots & a_{1n} \\ 0 & 0 & a_{23} & \dots & a_{2n} \\ 0 & 0 & 0 & \dots & a_{3n} \\ \vdots & & & & \vdots \\ 0 & 0 & 0 & \dots & 0 \end{bmatrix}$$

In symbolic form, this can be expressed as

$$\mathbf{A} = \mathbf{D}+\mathbf{L}+\mathbf{U}$$

which makes it possible to rewrite the equation system as

$$\mathbf{D}\boldsymbol{x} = -(\mathbf{L}+\mathbf{U})\boldsymbol{x}+\boldsymbol{b}$$

or, in the original notation

$$\begin{aligned} a_{11}x_1 &= -(\quad\quad a_{12}x_2+a_{13}x_3+ \dots +a_{1n}x_n) \quad +b_1 \\ a_{22}x_2 &= -(a_{21}x_1 \quad\quad +a_{23}x_3+ \dots +a_{2n}x_n) \quad +b_2 \\ a_{33}x_3 &= -(a_{31}x_1+a_{32}x_2+ \quad \dots +a_{3}x_{3n}) \quad +b_3 \\ &\vdots \\ a_{nn}x_n &= -(a_{n1}x_1+a_{n2}x_2+a_{n3}x_3+ \dots +a_{nn-1}x_{n-1})+b_n \end{aligned}$$

Note that an arbitrarily chosen, jth equation permits the calculation of x_j as a function of all the remaining variables. In fact, this is the underlying principle of the iteration formula which can be written in the original notation as

$$\begin{aligned} x_1^{(i+1)} &= c_{11}x_1^{(i)}+c_{12}x_2^{(i)}+ \dots +c_{1n}x_n^{(i)}+d_1 \\ x_2^{(i+1)} &= c_{21}x_1^{(i)}+c_{22}x_2^{(i)}+ \dots +c_{2n}x_n^{(i)}+d_2 \\ &\vdots \\ x_n^{(i+1)} &= c_{n1}x_1^{(i)}+c_{n2}x_2^{(i)}+ \dots +c_{nn}x_n^{(i)}+d_n \end{aligned}$$

or, in matrix notation

$$\boldsymbol{x}^{(i+1)} = \mathbf{C}\boldsymbol{x}^{(i)}+\boldsymbol{d}$$

The convergence of the Gauss–Seidel method, as well as its accuracy, depend on the properties of matrix **C**. If the entries in this matrix satisfy one of the conditions

$$\sum_{j=1}^{n} |c_{ij}| \leqslant w < 1 \qquad i = 1, 2, \ldots, n$$

$$\sum_{i=1}^{n} |c_{ij}| \leqslant v < 1 \qquad j = 1, 2, \ldots, n$$

then the iterative process converges to a solution from arbitrarily chosen vector of initial approximations $\boldsymbol{x}^{(0)}$.

The error of the solution can be estimated as follows:

– if the first condition is satisfied, then

$$|x_k - x_k^{(i)}| \leqslant \frac{w}{1-w} \max_{j=1,2,\ldots,n} |x_j^{(i)} - x_j^{(i-1)}|$$

– or, if the second condition is satisfied, then

$$|x_k - x_k^{(i)}| \leqslant \frac{v}{1-v} \sum_{j=1}^{n} |x_j^{(i)} - x_j^{(i-1)}|$$

Let us consider now a system of nonlinear equations in many unknowns, assuming that its general form is

$$\begin{aligned} g_1(x_1, x_2, x_3, \ldots, x_n) &= 0 \\ g_2(x_1, x_2, x_3, \ldots, x_n) &= 0 \\ g_3(x_1, x_2, x_3, \ldots, x_n) &= 0 \\ \vdots \qquad & \quad \vdots \\ g_n(x_1, x_2, x_3, \ldots, x_n) &= 0 \end{aligned} \qquad \text{(A.6)}$$

where: $g_1, g_2, \ldots, g_n$ – certain nonlinear functions.

The Gauss–Seidel method can be employed in solving equation system (A.6) only if it could be transformed to

$$\begin{aligned} x_1 &= G_1(\quad x_2, x_3, \ldots, x_n) \\ x_2 &= G_2(x_1, \quad x_3, \ldots, x_n) \\ x_3 &= G_3(x_1, x_2, \quad \ldots, x_n) \\ \vdots \\ x_n &= G_n(x_1, x_2, x_3, \ldots, x_{n-1}) \end{aligned}$$

where: $G_1, G_2, \ldots, G_n$ are also nonlinear functions. The iteration process is organized in exactly the same manner as in the case of systems of linear equations. It might happen in the nonlinear case, however, that convergence is difficult or even impossible to obtain.

It should be added that one of the most efficient techniques used in solving systems of nonlinear equations consists of forming a function

$$W(x) = [g_1(x)]^2 + [g_2(x)]^2 + \ldots + [g_n(x)]^2$$

and subsequently finding its minimum (the function value at the minimum is zero). This is usually done by the numerical methods of unconditional minimization.

Example A.4

Apply the Gauss–Seidel method to solve, with an accuracy of 0.01, the system of equations

$$\begin{cases} 0.96x_1+0.48x_2^2-0.12x_3^2 = 0.96 \\ 0.51x_1^2+1.02x_2-0.17x_3 = 1.53 \\ 0.10x_1+0.2x_2^2 - \quad x_3 = 1.10 \end{cases}$$

The system can be transformed into

$$\begin{cases} x_1 = 1.00-0.5x_2^2+0.12x_3^2 \\ x_2 = 1.50-0.5x_1^2+0.17x_3 \\ x_3 = -1.10+0.1x_1+0.2x_2^2 \end{cases}$$

As an initial approximation of a solution, (0, 0, 0) is assumed. Consecutive approximations are given in Table A.5.

Table A.5

Iterative approximations of a solution to equation system in Example A.4

i	1	2	3	4	5	6	7
x_1	1	0.58	0.51	0.48	0.46	0.44	0.44
x_2	1	1.07	1.10	1.12	1.13	1.13	1.13
x_3	−0.8	−0.81	−0.81	−0.81	−0.81	−0.81	−0.81

Example A.5

Solve, with an accuracy of 0.01, the system of equations

$$0.2x+0.3y+1.1z = \sqrt{1+x^2}$$

$$1.3x-0.2y-0.5z = \sqrt{1+y^2}$$

$$0.1x+1.1y-0.4z = \sqrt{1+z^2}$$

Let us change the order of equations and rewrite them as follows

$$1.3x-\left(0.2y+\sqrt{1+y^2}\right)-0.5z = 0$$

$$0.1x+1.1y+\left(0.4z-\sqrt{1+z^2}\right) = 0$$

$$\left(0.2x-\sqrt{1+x^2}\right)+0.3y+1.1z = 0$$

Now, two different solution procedures can be employed.

Procedure I. The system of equations can be transformed into

$$x = 0.154y + \frac{1}{1.3}\sqrt{1+y^2} + 0.385z$$

$$y = -0.091x - 0.364z + \frac{1}{1.1}\sqrt{1+z^2}$$

$$z = -0.182x + \frac{1}{1.1}\sqrt{1+x^2} - 0.273y$$

This makes it possible to solve the system by the Gauss–Seidel method. Assuming initial approximation vector (1, 1, 1), the iteration process obtained is shown in Table A.6.

Table A.6

Iterative process according to Gauss–Seidel method, Example A.5

i	1	2	3	4	5
x	1.63	1.56	1.59	1.59	1.59
y	0.77	0.85	0.83	0.84	0.84
z	1.23	1.17	1.19	1.19	1.19

Procedure II. Let us apply a relaxation technique in which errors of the equations are used for improving the approximation of a solution. If the current approximation is $(x^{(i)}, y^{(i)}, z^{(i)})$ then the respective errors are

$$R_1^{(i)} = 1.3x^{(i)} - \left(0.2y^{(i)} + \sqrt{1+x^{(i)2}}\right) - 0.5z^{(i)}$$

$$R_2^{(i)} = 0.1x^{(i)} + 1.1y^{(i)} + \left(0.4z^{(i)} - \sqrt{1+z^{(i)2}}\right)$$

$$R_3^{(i)} = \left(0.2x^{(i)} - \sqrt{1+x^{(i)2}}\right) + 0.3y^{(i)} + 1.1z^{(i)}$$

If the error vector is known at every step of the Gauss–Seidel iterative process, then a correction can be introduced into solution vector X according to the formula

$$X^{*(i)} = X^{(i)} + Y^{(i)}$$

where: $Y^{(i)}$ – correction vector with the components

$$Y_1^{(i)} = \frac{1}{a_x} R_1^{(i)}$$

$$Y_2^{(i)} = \frac{1}{a_y} R_2^{(i)}$$

$$Y_3^{(i)} = \frac{1}{a_z} R_3^{(i)}$$

This technique improves the convergence of the Gauss–Seidel method, when the values of coefficients a_x, a_y, a_z are chosen appropriately, as demonstrated below.

Let $a_x = 1.3$, $a_y = 1.1$, $a_z = 1.1$ and assume the same initial approximation as above. The resulting iteration process is shown in Table A.7. Note that the calculation workload exceeds that of the simple Gauss–Seidel procedure. Now, let $a_x = -1.3$, $a_y = -1.1$, $a_z = -1.1$; the iteration process shown in Table A.8 can be terminated at the third step, while five steps were necessary for the simple Gauss–Seidel procedure.

Table A.7

Iterative process of solving a system of equations (the first relaxation attempt)

i	1	2	3	4
x	1.63	1.51	1.63	1.59
y	0.77	0.85	0.84	0.84
z	1.23	1.13	1.21	1.19
R_1	0.08	−0.09	0.03	0
R_2	−0.08	0.03	0.005	0
R_3	0.002	0	0	0
x^*	1.69	1.44	1.65	
y^*	0.69	0.88	0.84	
z^*	1.23	1.23	1.21	

Table A.8

Iterative process of solving a system of equations (the second relaxation attempt)

i	1	2	3
x	1.63	1.60	1.59
y	0.77	0.84	0.84
z	1.23	1.19	1.19
R_1	0.08	0.01	0
R_2	−0.08	0	0
R_3	0.002	0	0
x^*	1.57	1.59	
y^*	0.84	0.84	
z^*	1.23	1.19	

As a rule, the application of the relaxation technique to large systems of equations results in considerable economy in the number of operations needed to arrive at the solution. Relaxation sometimes makes it possible to achieve convergence even if the simple iteration process is not convergent.

APPENDIX 4

Pressure losses in double-pipe segments at Reynolds numbers above 10000

Below, an explanation is given on the use of the nomograph shown in Fig. A.7, from which pressure losses in liquids flowing through a double-pipe segment can be determined.

(1) Determination of the pressure loss in the inner pipe (complete with a return bend) of a single segment, Δp_1 N/m².

This can be done with the aid of diagrams 1 and 2. The procedure is as follows:

(1.1) Find Reynolds number Re_1 on axis A in diagram 1.

(1.2) Calculate $\phi_1 = \dfrac{L}{2(d_1 - 2g_1)} + 25$ and find y_1 on axis C in diagram 1.

(1.3) Calculate $\dfrac{u_1 \mu_1}{d_1}$ and find its value on axis E in diagram 2.

(1.4) Draw a line between points found on axes C and E; read Δp_1 on axis D in diagram 1.

Example. Determine pressure loss in the inner pipe together with the return bend. Pipe diameter is $d_1 = 57$ mm, wall thickness $g_1 = 3$ mm, segment length $L = 3$ mm. Liquid flow is characterized by $Re_1 = 10^5$, $u_1 = 1.7$ m/s and $\mu_1 = 10^{-4}$ N s/m².

The solution:

$$\phi_1 = \frac{3}{2(0.057 - 2 \times 0.003)} + 25 = 57.5$$

From diagram 1, we find $y_1 = 1.05 \times 10^5$.

$$\frac{u_1 \mu_1}{d_1} = \frac{1.7 \times 10^{-4}}{0.057} = 3 \times 10^{-3} \ \text{N/m}^2$$

From diagram 2, we find $\Delta p_1 = 3 \times 10^2$ N/m².

(2) Determining pressure loss in the outer channel (complete with nozzles) of a single segment, Δp_2 N/m².

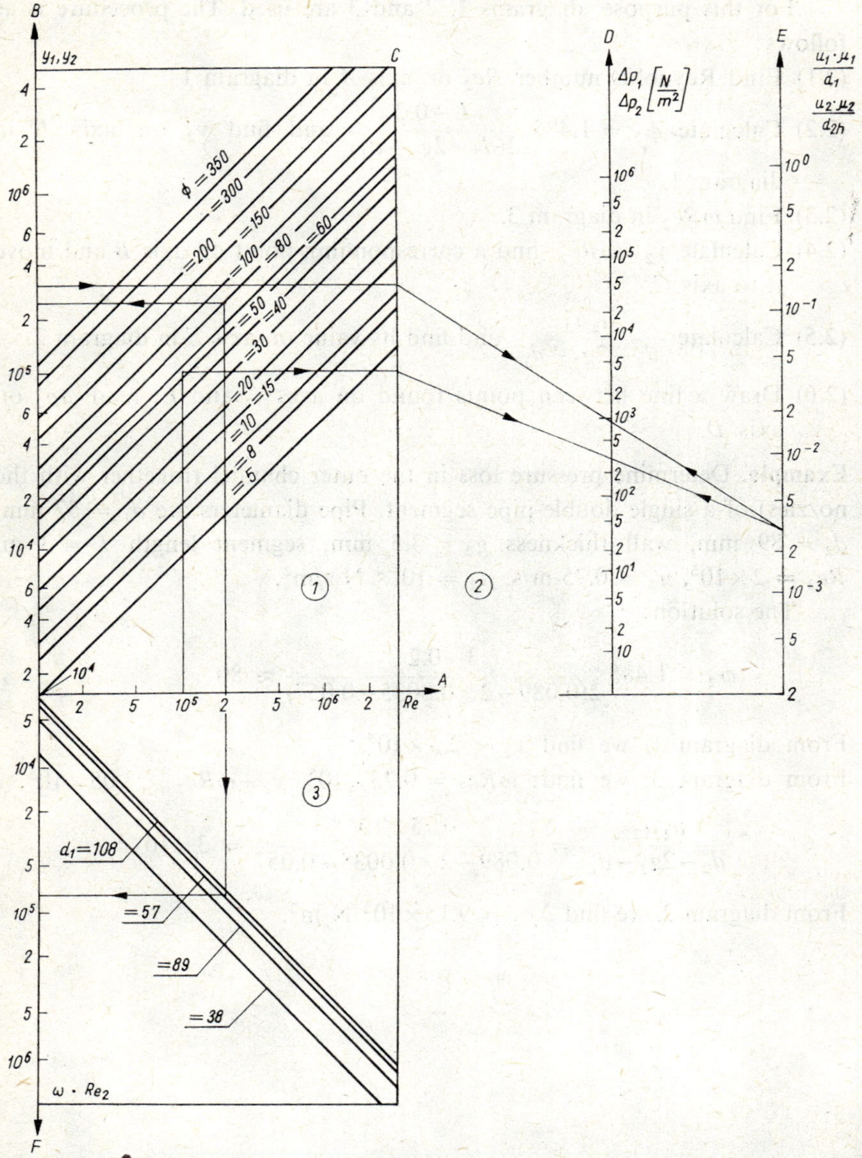

Fig. A.7 Nomograph for determining pressure losses in double-pipe segments.

Symbols: $d_{2h} = (d_2 - 2g_2) - d_1$; $\phi_1 = \dfrac{L}{2(d_1 - 2g_1)} + 25$; $\phi_2 = 1.485 \dfrac{L - 0.2}{2[(d_2 - 2g_2) - d_1]}$;

$y_1 = 0.184 \cdot \phi_1 Re_1^{0.8}$; $\quad y_2 = 0.184 \cdot \phi_2 \, Re_2^{0.8} + \omega \cdot Re_2$;

$$\omega = 0.364 + \left[\sqrt{2}\,\frac{(d_2 - dg_2)^2 - d_1^2}{(d_r - 2g_r)^2} - 1\right]^2 + \frac{2}{3}\left[1 - \frac{\sqrt{2}}{2}\,\frac{(d_r - 2g_r)^2}{(d_2 - 2g_2)^2 - d_1^2}\right];$$

d_r – inner diameter of the return bend.

For this purpose, diagrams 1, 2 and 3 are used. The procedure is as follows:

(2.1) Find Reynolds number Re_2 on axis A in diagram 1.

(2.2) Calculate $\phi_2 = 1.485 \dfrac{L-0.2}{2(d_2-2g_2-d_1)}$ and find y_2 on axis B in diagram 1.

(2.3) Find ωRe_2 in diagram 3.

(2.4) Calculate $y_2+\omega Re_2$, find a corresponding point on axis B and move it to axis C.

(2.5) Calculate $\dfrac{u_2\mu_2}{d_2-2g_2-d_1}$ and find its value on axis E in diagram 2.

(2.6) Draw a line between points found on axes C and E; read Δp_2 on axis D.

Example. Determine pressure loss in the outer channel (together with the nozzles) of a single double-pipe segment. Pipe diameters are $d_1 = 57$ mm, $d_2 = 89$ mm, wall thickness $g_2 = 3.5$ mm, segment length $L = 3$ m, $Re_2 = 2\times10^5$, $u_2 = 0.75$ m/s, $\mu_2 = 10^{-4}$ N s/m².

The solution:

$$\phi_2 = 1.485 \frac{3-0.2}{2(0.089-2\times0.0035-0.057)} \approx 80$$

From diagram 1, we find $y_2 = 2.3\times10^5$.
From diagram 3, we find: $\omega Re_2 = 0.75\times10^5$, $y_2+\omega Re_2 = 3.05\times10^5$.

$$\frac{u_2\mu_2}{d_2-2g_2-d_1} = \frac{0.75\times10^{-4}}{0.089-2\times0.0035-0.057} = 3\times10^{-3}$$

From diagram 2, we find $\Delta p_2 = 9.15\times10^2$ N/m².

References

[1] Afimiwala, K. A., Mayne, R. W., Evaluation of Optimization Techniques for Applications in Engineering Design, *J. Spacecr. Rockets*, No. 10, 1974.

[2] Asimov, M., *Introduction to Engineering Design*, McGraw-Hill, New York, 1966.

[3] Avriel, M., Rijckaert, M. J., Wilde, D. J., *Optimization and Design*, Prentice Hall, Englewood Cliffs, 1973.

[4] Bator, G., Urbaniec, K., Projektierung von Verdampfanlagen in Zuckerfabriken mit Hilfe von Computern, *Zuckerindustrie*, No. 12, 1978.

[5] Beckenbach, E. F. (ed.), *Modern Mathematics for the Engineer*, McGraw-Hill, New York, 1956.

[6] Beightler Ch. S., Phillips D. T., Wilde D. J., *Foundations of Optimization*, Prentice Hall, Englewood Cliffs, 1979.

[7] Bekey, G. A., Ung, M. T., A Comparative Evaluation of Two Global Search Algorithms, *IEEE Trans. Systems, Man and Cybernetics*, January 1974.

[8] Brandt, A. M. (ed.), *Criteria and Methods for Optimization of Structures* (in Polish), PWN, Warsaw, 1977.

[9] Brownell, L. E., Young, E. H., *Process Equipment Design — Vessel Design*, Wiley, New York, 1959.

[10] Boyarinov, A. I., Kafarov, V. V., *Optimization Methods in Chemical Engineering* (in Russian), Khimija, Moscow, 1970.

[11] Chawla, J. M., Wärmeübergang und Druckverlust bei der Kältemittelverdampfung im waagerechten Strömungsrohr, *Chem.-Ing. Technik*, No. 5, 1968.

[12] Conley, W., *Computer Optimization Techniques*, Petrocelli Books, New York, 1980.

[13] Eason, E. D., Fenton, R. G., A Comparison of Numerical Optimization Methods for Engineering Design, *Trans. ASME, J. Engng for Industry*, February 1974.

[14] Eder, H., Dimensioning of Fins with General Nonlinear Cooling Laws, *Wärme- und Stoffübergang*, No. 2, 1973.

[15] Fiacco, A. V., Mc Cormick, G. P., *Nonlinear Programming: Sequential Unconstrained Minimization Techniques*, Wiley, New York, 1968.

[16] Findeisen, W., Szymanowski, J., Wierzbicki, A., *Theory and Computational Methods for Optimization* (in Polish), PWN, Warsaw, 1977.

[17] Fraenkel, A. A., Bar-Hillel, Y., Levy, A., *Foundations of Set Theory*, North-Holland, Amsterdam-London, 1973.

[18] Ginsberg, M., A Guide to the Literature of Modern Numerical Mathematics, *Comp. Rev.*, No. 2, 1975.

[19] Gass, S. I., *Linear Programming. Methods and Applications*, McGraw-Hill, New York, 1958.

[20] Gerrard, A. M., Liddle, C. J., The Optimal Selection of Multiple Hydrocyclone Systems, *Chemical Engineer*, No. 297, 1975.

[21] Gisvold, K. M., Moe, J., A Method for Nonlinear Mixed-Integer Programming and Its Application to Design Problems, *Trans. ASME, J. Engng for Industry*, May 1972.

[22] Goliński, J., *Optimization Methods for Technical Design* (in Polish), WNT, Warsaw, 1974.

[23] Gostyunin, A. G., Criteria of Economical Shape of Tube-in-Shell Heat Exchangers (in Russian), *Khim. i Neft. Machinostr.*, No. 3, 1974.

[24] Greenberg, H. J., Loh, L. W. T., An Analysis of Cut-off Rules for Optimization Algorithms, *IEEE Trans. Systems, Man and Cybernetics*, January 1974.

[25] Gutkowski, K., *Refrigeration. Selected Computational Problems* (in Polish), WNT, Warsaw, 1972.

[26] Himmelblau, D. M., *Applied Nonlinear Programming*, McGraw-Hill, New York, 1972.

[27] Himmelblau, D. M. (ed.), *Decomposition of Large-Scale Problems*, North-Holland, Amsterdam, 1973.

[28] Hoffmann, U., Hofmann, H., *Einführung in die Optimierung mit Anwendungsbeispielen aus dem Chemie–Ingenieur–Wesen*, Verlag Chemie, Weinheim, 1971.

[29] Hommann, G., Optimale Auslegung von Wärmeübertragern mit Verdampfung in horizontalen Rohren, *Luft- und Kältetechnik*, No. 1, 1974.

[30] Johnson R. C., *Optimum Design of Mechanical Elements*, 2nd ed., Wiley, New York, 1980.

[31] Johnson R. C., *Mechanical Design Synthesis with Optimization Applications*, Van Nostrand Reinhold, New York, 1971.

[32] Kafarov, V. V., *Cybernetical Methods in Chemistry and Chemical Engineering* (in Russian), Khimija, Moscow, 1976.

[33] Korbut, A. A., Finkelshtein, Y. Y., *Discrete Programming* (in Russian), Nauka, Moscow, 1969.

[34] Krick, E. V., *An Introduction to Engineering and Engineering Design*, Wiley, New York, 1969.

[35] Kriegel, E., Brauer, H., Hydraulischer Transport körniger Feststoffe durch waagerechte Rohrleitungen, *VDI-Forschung, Heft* 515, 1966.

[36] Kubasiewicz, A., Criteria for Selection of Compact Heat Exchangers (in Polish), *Inż. i Ap. Chem.*, No. 4, 1972.

[37] Kubasiewicz, A., *et al.*, Optimum Design of Thermal Systems of Sugar Plants, *Proc. 8th Int. Conf. on Industrial Energetics*, Gdańsk, 1975.

[38] Kubasiewicz, A., Urbaniec, K., Methodology of Design Standardization in Process Equipment Industry (in Polish), *Inż. i Ap. Chem.*, No. 5, 1976.

[39] Kubarski J., Urbaniec K., Optimum Series of Condensate Tanks (in Polish), *Gaz. Cukrown.*, No. 7, 1977.

[40] Kuester, J. L., Mize, J. H., *Optimization Techniques with FORTRAN*, McGraw-Hill, New York, 1973.

[41] Kuratowski, K., Mostowski, A., *Set Theory*, PWN, Warsaw, 1968.

[42] Kühns, W., Zur kostengünstigen Auslegung von luftgekühlten Kondensatoren, *Verfahrenstechnik*, No. 6, 1973.

[43] Land, A., Powell, S., *FORTRAN Codes for Mathematical Programming: Linear, Quadratic and Discrete*, Wiley-Interscience, Chichester, 1973.

[44] Lazarev, I. B., Valuyskikh, V. P., Experimental Comparison of Random Search Algorithms (in Russian), *Avtom. i Vycisl. Tekh.*, No. 2, 1974.

[45] Leaman, J. M., A Simple Way to Optimize, *Machine Design*, March 21, 1974.

[46] Littlefield, J. M., Cox, J. E., Optimization of Annular Fins on a Horizontal Tube, *Wärme- und Stoffübergang*, No. 2, 1974.

[47] Łajkowski, J., Optimization of Standard Heat Exchangers (in Polish), Ph. D. Dissertation, Warsaw Technical University, 1976.

[48] Maňas, M., *Optimization Methods* (in Czech), SNTL, Prague, 1979.

[49] Mangasarian, O. L., Techniques of Optimization, *Trans. ASME, J. Engng for Industry*, May 1972.

[50] Mathematical Programming System – IBM Program Package 360 A – Co. 14X.

[51] McCormick, G. P., Algorithmic and Computational Aspects of the Use of Optimization Methods in Engineering Design, *Comp. and Structures*, No. 6, 1973.

[52] Mischke, C. R., *An Introduction to Computer-Aided Design*, Prentice Hall, Englewood Cliffs, 1968.

[53] Oderfeld, J., Optimal Synthesis of Machines (in Polish), Collected Papers from the Dept. of Theory of Machines and Mechanisms, Warsaw Technical University, 1966.

[54] Opalińska, K., Siarkiewicz, M., Urbaniec, K., Computerized Design of Fired Heaters for Petrochemical Installations, *Proc. 3rd Symp. Use of Computers in Chemical Engineering*, Gliwice, 1974.

[55] Ostrovskij, G. M., Volin Yu. M., *Methods for Optimization of Complex Systems of Chemical Technology* (in Russian), Khimija, Moscow 1970.

[56] Pappas, M., Allentuch, A., Mathematical Programming Procedures for Mixed Discrete-Continuous Design Problems, *Trans. ASME, J. Engng for Industry*, February 1974.

[57] Pho, T. K., Lapidus, L., Topics in Computer-Aided Design, *AIChE J.*, No. 6, 1973.

[58] Peters, D. L., Nicole, F. J. L., Efficient Programming for Cost-Optimised Heat Exchanger Design, *Chemical Engineer*, No. 259, 1972.

[59] Pinter, C. C., *Set Theory*, Addison-Wesley, Reading, 1971.

[60] Pierre, D. A., *Optimization Theory with Applications*, Wiley, New York, 1969.

[61] Pervozvanskij, A. A., *Search Methods* (in Russian), Nauka, Moscow, 1970.

[62] Pogorzelski, W., *Optimization of Technical Systems in Examples* (in Polish), WNT, Warsaw, 1978.

[63] Price, W. L., A Controlled Random Search Procedure for Global Optimization, *Computer J.*, No. 4, 1977.

[64] Ralston, A., *A First Course in Numerical Analysis*, McGraw-Hill, New York, 1965.

[65] Rastrigin, L. A., *Extremal Control Systems* (in Russian), Nauka, Moscow, 1974.

[66] Rietema, K., Performance and Design of Hydrocyclones, *Chemical Engineering Science*, No. 3/4, 1961.

[67] Rorres C., Anton H., *Applications of Linear Algebra*, Wiley, Chichester, 1977.

[68] Rudd, D. F., Watson, C. C., *Strategy of Process Engineering*, Wiley, New York, 1968.

[69] Rutkowski, L., Urbaniec, K., Automated Computations of Fired Heaters for Thermal and Hydraulic Design (in Polish), *Inż. i Ap. Chem.*, No. 3, 1976.

[70] Rutkowski, L., Urbaniec, K., Zgorzelski, M., Optimization of Double-Pipe Heat Exchangers (in Polish), *Inż. i Ap. Chem.*, No. 1, 1972.

[71] Sears J. L., *Optimization Techniques in FORTRAN*, Petrocelli Books, New York, 1979.

[72] Seireg, A., A Survey of Optimization of Mechanical Design, *Trans. ASME, J. Engng for Industry*, May 1972.

[73] Shevchenko, V. S., *Introduction to Optimal Design of Machines* (in Russian), Nauka i Tekhnika, Minsk, 1974.

[74] Shnell, R. V., Application of Game Theory to Solution Finding in Power Engineering Problems Under Uncertainty Conditions (in Russian), *Izv. AN SSSR*, Energetics and Transportation, No. 6, 1972.

[75] Slipcevič, B., Graphische Optimierung von Kühlern mit Verdampfung von Kältemitteln in horizontalen Rohren, *Verfahrenstechnik*, No. 1, 1972.

[76] *Thermal Handbook*, Alfa-Laval (Sweden), 1969.

[77] Urbaniec, K., Simplification of Series of Heat Exchangers – Optimizing the Solution for Double-Pipe Heat Exchangers, *Proc. 2nd Symp. Applications of Mathematics to Standardization*, Warsaw, 1972.

[78] Urbaniec, K., Optimization of the Type Series, *Operational Research Quarterly*, No. 4 ii, 1976.

[79] Urbaniec, K., Optimization of Series of Process Equipment (in Polish), *Inż. i Ap. Chem.*, No. 5, 1976.

[80] Urbaniec, K., Applied Optimization – a Review, Paper presented at UNIDO Interregional Expert Group Meeting on Computer Applications and Engineering, Warsaw, 1977 (unpublished).

[81] Urbaniec, K., Equipment for the Thermal Systems of Sugar Factories, *Polish Techn. Rev.*, No. 7, 1979.

[82] Urbaniec, K., *et al.*, New Series of Evaporators (in Polish), *Gaz. Cukrown.*, No. 3, 1978.

[83] Urbaniec, K., Kulig, M., Optimal Design of a Series of Heat Exchangers, *Proc. 5th Symp. Computers in Chemical Engineering*, Vysoke Tatry, 1977.

[84] Urbaniec, K., Siarkiewicz, M., Optimal Design of Fired Heaters, *Proc. Symp. Computers in Design and Erection of Chemical Plants*, Karlovy Vary, 1975.

[85] Urbaniec, K., Zgorzelski, M., Optimization of Thermal Equipment and Systems (in Polish), Paper presented at the Symposium 'Optymalizacja w Mechanice', Beskid Śląski, 1976 (unpublished).

[86] Warren, A. D., The Status of Nonlinear Programming Software, *Oper. Res.*, No. 3, 1979.

[87] Wilde, D. J., The Monotonicity Table in Optimal Engineering Design, *Engng Optimization*, No. 1, 1976.

[88] Volgin, L. N., *Optimization Problems in Theoretical Cybernetics* (in Russian), Sov. Radio, Moscow, 1968.

[89] Zakharov, V. V., Ten Widely Used Test Functions for Optimization Methods (in Russian), *Avtom. i Vycisl. Tekhn.*, No. 6, 1974.

[90] Zangwill, W. I., *Nonlinear Programming, a Unified Approach*, Prentice Hall, Englewood Cliffs, 1969.

[91] Zener, C., *Engineering Design by Geometric Programming*, Wiley-Interscience, New York, 1971.

Index